Quantum Monte Carlo

Quantum Monte Carlo

Origins, Development, Applications

James B. Anderson

Department of Chemistry
Department of Physics
The Pennsylvania State University

UNIVERSITY PRESS
2007

OXFORD
UNIVERSITY PRESS

Oxford University Press, Inc., publishes works that further
Oxford University's objective of excellence
in research, scholarship, and education.

Oxford   New York
Auckland   Cape Town   Dar es Salaam   Hong Kong   Karachi
Kuala Lumpur   Madrid   Melbourne   Mexico City   Nairobi
New Delhi   Shanghai   Taipei   Toronto

With offices in
Argentina   Austria   Brazil   Chile   Czech Republic   France   Greece
Guatemala   Hungary   Italy   Japan   Poland   Portugal   Singapore
South Korea   Switzerland   Thailand   Turkey   Ukraine   Vietnam

Copyright © 2007 by Oxford University Press

Published by Oxford University Press, Inc.
198 Madison Avenue, New York, New York 10016

www.oup.com

Oxford is a registered trademark of Oxford University Press

Library of Congress Cataloging-in-Publication Data
Quantum Monte Carlo : origins, development, applications /
by James B. Anderson.
     p. cm.
   Includes index.
   ISBN 978-0-19-531010-8
   1. Monte Carlo method—Abstracts.   2. Quantum theory—Abstracts.
I. Anderson, James B., 1935–
QC20.7.M65Q37 2006
530.1201′518282—dc22   2006040145

9 8 7 6 5 4 3 2 1
Printed in the United States of America
on acid-free paper

To Nancy

. . . many shall run to and fro,
and knowledge shall be increased.

Daniel 12:4

PREFACE

In studying quantum mechanics and especially quantum chemistry I found a little book written by Henry F. Schaefer III to be especially useful in providing an overview of the subject, a guide to its trends, and an easily researched compendium of important papers in the area. In writing this new book in the area of quantum Monte Carlo (QMC) I hope to have provided the same sorts of services to those studying and working in the area of quantum Monte Carlo methods.

Schaefer's book is titled *Quantum Chemistry* and subtitled *The development of ab initio methods in molecular electronic structure theory* [Oxford University Press, 1984]. In adopting a similar approach to a review of QMC I hope that Professor Schaefer considers my following his example as flattery.

The term "Monte Carlo" was first used to describe calculational methods based on chance in the 1940s, but the methods themselves preceded the term by as much as a century. The combination of "quantum" and "Monte Carlo" first appeared in 1981 and, similarly, was preceded by early developments of the methods.

In this book I have attempted to collect summaries of some of the most important papers in the quantum Monte Carlo literature. In doing so I have chosen those papers which describe the bases for the many different methods within QMC, provide samples of the works of many of those responsible for the development of QMC, and illustrate the breadth of applications of QMC. The selection is, of course, somewhat random, somewhat arbitrary, and surely arguable. I express my apologies to those authors whose papers may be underrepresented.

The success of QMC methods over the past few decades has been remarkable, and the papers described in this book clearly demonstrate that success. For isolated molecules, the basic material of chemistry, QMC methods have produced "exact" solutions of the Schrödinger equation for very small systems and the most accurate solutions available for very large systems. The range of applications is impressive for other types of problems: folding of protein molecules, interactions in liquids, band structures in crystals, quantum dots, enzyme structure, energetics of light harvesting materials, potential energy surfaces for reactions of all types, transition dipole moments, structure of a water droplet, frequency shifts in heteroclusters, energetics of excitons, binding in nuclei, and interdimensional degeneracies. Many of these are new problems, approached successfully for the first time using QMC methods.

One may certainly expect the further development of QMC methods and many more applications. It is my hope that this book will in some small way provide assistance to those engaged in the effort.

I am happy to acknowledge the assistance of Dr. Nathan M. Urban in producing the manuscript and of Professor Stuart M. Rothstein and Professor Arne Lüchow for their suggestions and corrections.

An Incomplete List of Important Papers in Quantum Monte Carlo

1

E. Schrödinger
Über die Umkehrung der Naturgesetze
Sitzber. Preuss. Akad. Wiss. Phys.-math. Kl., 144–153 (1931)

2

N. Metropolis & S. Ulam
The Monte Carlo method
J. Am. Stat. Assoc. **44**, 335–341 (1949)

3

M. H. Kalos
Monte Carlo calculations of the ground state of three- and
    four-body nuclei
Phys. Rev. **128**, 1791–1795 (1962)

4

H. Conroy
Molecular Schrödinger equation. II. Monte Carlo evaluation of
    integrals
J. Chem. Phys. **41**, 1331–1335 (1964)

5

W. L. McMillan
Ground state of liquid ^{4}He
Phys. Rev. **138**, A442–A451 (1965)

6

M. H. Kalos
Stochastic wave function for atomic helium
J. Comput. Phys. **1**, 257–276 (1967)

7

R. C. Grimm & R. G. Storer
Monte-Carlo solution of Schrödinger's equation
J. Comput. Phys. **7**, 134–156 (1971)

8

M. H. Kalos, D. Levesque, & L. Verlet
Helium at zero temperature with hard-sphere
and other forces
Phys. Rev. A **9**, 2178–2195 (1974)

9

K. S. Liu, M. H. Kalos, & G. V. Chester
Quantum hard spheres in a channel
Phys. Rev. A **10**, 303–308 (1974)

10

J. B. Anderson
A random-walk simulation of the Schrödinger equation: H_3^+
J. Chem. Phys. **63**, 1499–1503 (1975)

11

D. J. Klein & H. M. Pickett
Nodal hypersurfaces and Anderson's random-walk simulation of
the Schrödinger equation
J. Chem. Phys. **64**, 4811–4812 (1976)

12

J. B. Anderson
Quantum chemistry by random walk. H ^2P, H_3^+ D_{3h} $^1A_1'$,
H_2 $^3\Sigma_u^+$, H_4 $^1\Sigma_g^+$, Be ^1S
J. Chem. Phys. **65**, 4121–4127 (1976)

13

R. L. Coldwell & R. E. Lowther
Monte Carlo calculation of the Born-Oppenheimer potential
between two helium atoms using Hylleraas-type electronic
wave functions
Int. J. Quantum Chem., Symp. Ser. **12**, 329–341 (1978)

14
J. B. Anderson
Quantum chemistry by random walk: H_4 square
Int. J. Quantum Chem. **15**, 109–120 (1979)

15
J. B. Anderson & B. H. Freihaut
Quantum chemistry by random walk: Method of successive
  corrections
J. Comput. Phys. **31**, 425–437 (1979)

16
Y. Tomashima & J. Ozaki
Monte Carlo solution of Schrödinger's equation for the hydrogen
  atom in a magnetic field
J. Comput. Phys. **33**, 382–396 (1979)

17
J. B. Anderson
Quantum chemistry by random walk: Higher accuracy
J. Chem. Phys. **73**, 3897–3899 (1980)

18
D. M. Ceperley & B. J. Alder
Ground state of the electron gas by a stochastic method
Phys. Rev. Lett. **45**, 566–569 (1980)

19
F. Mentch & J. B. Anderson
Quantum chemistry by random walk: Importance sampling for H_3^+
J. Chem. Phys. **74**, 6307–6311 (1981)

20
K. McDowell & J. D. Doll
Quantum Monte Carlo and the hydride ion
Chem. Phys. Lett. **81**, 335–338 (1981)

21
K. McDowell
Assessing the quality of a wavefunction using quantum
  Monte Carlo
Int. J. Quantum Chem., Symp. Ser. **15**, 177–181 (1981)

22
J. G. Zabolitzky & M. H. Kalos
Solution of the four-nucleon Schrödinger equation
Nuc. Phys. A **356**, 114–128 (1981)

23
D. M. Arnow, M. H. Kalos, M. A. Lee, & K. E. Schmidt
Green's function Monte Carlo for few fermion problems
J. Chem. Phys. **77**, 5562–5572 (1982)

24
J. W. Moskowitz, K. E. Schmidt, M. A. Lee, & M. H. Kalos
Monte Carlo variational study of Be: A survey of
    correlated wavefunctions
J. Chem. Phys. **76**, 1064–1067 (1982)

25
J. W. Moskowitz, K. E. Schmidt, M. A. Lee, & M. H. Kalos
A new look at correlation energy in atomic and molecular systems.
    II. The application of the Green's function Monte Carlo
    method to LiH
J. Chem. Phys. **77**, 349–355 (1982)

26
P. J. Reynolds, D. M. Ceperley, B. J. Alder, & W. A. Lester, Jr.
Fixed-node quantum Monte Carlo for molecules
J. Chem. Phys. **77**, 5593–5603 (1982)

27
D. W. Heys & D. R. Stump
Application of the Green's-function Monte Carlo method to
    the compact Abelian lattice gauge theory
Phys. Rev. D **28**, 2067–2075 (1983)

28
V. R. Pandharipande, J. G. Zabolitzky, S. C. Pieper, R. B. Wiringa,
    & U. Helmbrecht
Calculations of ground-state properties of liquid ^{4}He droplets
Phys. Rev. Lett. **50**, 1676–1679 (1983)

29

M. A. Lee, P. Vashishta, & R. K. Kalia

Ground state of excitonic molecules by the Green's-function
Monte Carlo method

Phys. Rev. Lett. **51**, 2422–2425 (1983)

30

F. Mentch & J. B. Anderson

Quantum chemistry by random walk: Linear H_3

J. Chem. Phys. **80**, 2675–2680 (1984)

31

D. M. Ceperley & B. J. Alder

Quantum Monte Carlo for molecules: Green's function and
nodal release

J. Chem. Phys. **81**, 5833–5844 (1984)

32

P. J. Reynolds, R. N. Barnett, & W. A. Lester, Jr.

Quantum Monte Carlo study of the classical barrier height for
the $H + H_2$ exchange reaction: restricted versus unrestricted
trial functions

Int. J. Quantum Chem., Symp. Ser. **18**, 709–717 (1984)

33

R. K. Kalia, P. Vashishta, & M. A. Lee

Binding energy of positively charged acceptors in germanium —
A Green's function Monte Carlo calculation

Solid State Comm. **52**, 873–876 (1984)

34

P. J. Reynolds, M. Dupuis, & W. A. Lester, Jr.

Quantum Monte Carlo calculation of the singlet-triplet
splitting in methylene

J. Chem. Phys. **82**, 1983–1990 (1985)

35

J. B. Anderson

Quantum chemistry by random walk: A faster algorithm

J. Chem. Phys. **82**, 2662–2663 (1985)

36

D. F. Coker, R. E. Miller, & R. O. Watts
The infrared predissociation spectra of water clusters
J. Chem. Phys. **82**, 3554–3562 (1985)

37

D. Ceperley & B. J. Alder
Muon–alpha-particle sticking probability in muon-catalyzed fusion
Phys. Rev. A **31**, 1999–2004 (1985)

38

R. J. Harrison & N. C. Handy
Quantum Monte Carlo calculations on Be and LiH
Chem. Phys. Lett. **113**, 257–263 (1985)

39

B. H. Wells
The differential Green's function Monte Carlo method.
  The dipole moment of LiH
Chem. Phys. Lett. **115**, 89–94 (1985)

40

J. W. Moskowitz & K. E. Schmidt
The domain Green's function method
J. Chem. Phys. **85**, 2868–2874 (1986)

41

D. F. Coker & R. O. Watts
Quantum simulation of systems with nodal surfaces
Mol. Phys. **58**, 1113–1123 (1986)

42

P. J. Reynolds, R. N. Barnett, B. L. Hammond, R. M. Grimes, &
  W. A. Lester, Jr.
Quantum chemistry by quantum Monte Carlo: Beyond
  ground-state energy calculations
Int. J. Quantum Chem. **29**, 589–596 (1986)

43

G. Sugiyama & S. E. Koonin
Auxiliary field Monte Carlo for quantum many-body ground states
Ann. Phys. **168**, 1–26 (1986)

51
D. F. Coker & R. O. Watts
Diffusion Monte Carlo simulation of condensed systems
J. Chem. Phys. **86**, 5703–5707 (1987)

52
S. M. Rothstein, N. Patil, & J. Vrbik
Time step error in diffusion Monte Carlo simulations:
  An empirical study
J. Comput. Chem. **8**, 412–419 (1987)

53
G. J. Martyna & B. J. Berne
Structure and energetics of Xe_n^-
J. Chem. Phys. **88**, 4516–4525 (1988)

54
T. Yoshida & K. Iguchi
Quantum Monte Carlo method with the model potential
J. Chem. Phys. **88**, 1032–1034 (1988)

55
M. Caffarel & P. Claverie
Development of a pure diffusion quantum Monte Carlo method
  using a full generalized Feymann-Kac formula. I and II
J. Chem. Phys. **88**, 1088–1109 (1988)

56
D. R. Garmer & J. B. Anderson
Potential energies for the reaction $F + H_2 \rightarrow HF + H$ by
  the random walk method
J. Chem. Phys. **89**, 3050–3056 (1988)

57
S. Fahy, X. W. Wang, & S. G. Louie
Variational quantum Monte Carlo nonlocal pseudopotential
  approach to solids: Cohesive and structural properties of
  diamond
Phys. Rev. Lett. **61**, 1631–1634 (1988)

58

C. J. Umrigar, K. G. Wilson, & J. W. Wilkins
Optimized trial wave functions for quantum Monte Carlo
    calculations
Phys. Rev. Lett. **60**, 1719–1722 (1988)

59

B. L. Hammond, P. J. Reynolds, & W. A. Lester, Jr.
Damped-core quantum Monte Carlo method: Effective treatment
    for large-Z systems
Phys. Rev. Lett. **61**, 2312–2315 (1988)

60

J. Vrbik, M. F. DePasquale, & S. M. Rothstein
Estimating the relativistic energy by diffusion Monte Carlo
J. Chem. Phys. **88**, 3784–3787 (1988)

61

J. Carlson
Alpha particle structure
Phys. Rev. C **38**, 1879–1885 (1988)

62

C. A. Traynor & J. B. Anderson
Parallel Monte Carlo calculations to determine energy differences
    among similar molecular structures
Chem. Phys. Lett. **147**, 389–394 (1988)

63

M. Caffarel, P. Claverie, C. Mijoule, J. Andzelm, & D. R. Salahub
Quantum Monte Carlo method for some model and realistic
    coupled anharmonic oscillators
J. Chem. Phys. **90**, 990–1002 (1989)

64

J. Carlson, J. W. Moskowitz, & K. E. Schmidt
Model Hamiltonians for atomic and molecular systems
J. Chem. Phys. **90**, 1003–1006 (1989)

65

V. Dobrosavljević, C. W. Henebry, & R. M. Stratt

Simulation of the electronic structure of an atom dissolved in
a hard-sphere liquid

J. Chem. Phys. **91**, 2470–2478 (1989)

66

V. Mohan & J. B. Anderson

Effect of crystallite shape on exciton energy: Quantum
Monte Carlo calculations

Chem. Phys. Lett. **156**, 520–524 (1989)

67

G. Sugiyama, G. Zerah, & B. J. Alder

Ground-state properties of metallic lithium

Physica A **156**, 144–168 (1989)

68

H. Sun & R. O. Watts

Diffusion Monte Carlo simulations of hydrogen fluoride
dimers

J. Chem. Phys. **92**, 603–616 (1990)

69

V. Mohan & J. B. Anderson

Quantum Monte Carlo calculations of three-body corrections
in the interaction of three helium atoms

J. Chem. Phys. **92**, 6971–6973 (1990)

70

K. E. Schmidt & J. W. Moskowitz

Correlated Monte Carlo wave functions for the atoms
He through Ne

J. Chem. Phys. **93**, 4172–4178 (1990)

71

M. V. Rama Krishna & K. B. Whaley

Wave functions of helium clusters

J. Chem. Phys. **93**, 6738–6751 (1990)

72

S. Fahy, X. W. Wang, & S. G. Louie

Variational quantum Monte Carlo nonlocal pseudopotential
   approach to solids: Formulation and application to diamond,
   graphite, and silicon

Phys. Rev. B **42**, 3503–3522 (1990)

73

X. W. Wang, J. Zhu, S. G. Louie, & S. Fahy

Magnetic structure and equation of state of bcc solid hydrogen:
   A variational quantum Monte Carlo study

Phys. Rev. Lett. **65**, 2414–2417 (1990)

74

C. A. Traynor, J. B. Anderson, & B. M. Boghosian

A quantum Monte Carlo calculation of the ground state energy of
   the hydrogen molecule

J. Chem. Phys. **94**, 3657–3664 (1991)

75

M. Quack & M. A. Suhm

Potential energy surfaces, quasiadiabatic channels,
   rovibrational spectra, and intramolecular dynamics of
   $(HF)_2$ and its isotopomers from quantum Monte Carlo
   calculations

J. Chem. Phys. **95**, 28–59 (1991)

76

L. Mitáš, E. L. Shirley, & D. M. Ceperley

Nonlocal pseudopotentials and diffusion Monte Carlo

J. Chem. Phys. **95**, 3467–3475 (1991)

77

S. A. Alexander, R. L. Coldwell, H. J. Monkhorst, &
   J. D. Morgan III

Monte Carlo eigenvalue and variance estimates from several
   functional optimizations

J. Chem. Phys. **95**, 6622–6633 (1991)

78
J. B. Anderson, C. A. Traynor, & B. M. Boghosian
Quantum chemistry by random walk: Exact treatment of
    many-electron systems
J. Chem. Phys. **95**, 7418–7425 (1991)

79
G. An & J. M. J. Van Leeuwen
Fixed-node Monte Carlo study of the two-dimensional
    Hubbard model
Phys. Rev. B **44**, 9410–9417 (1991)

80
R. N. Barnett, P. J. Reynolds, & W. A. Lester, Jr.
Monte Carlo algorithms for expectation values of
    coordinate operators
J. Comput. Phys. **96**, 258–276 (1991)

81
M. Caffarel & O. Hess
Quantum Monte Carlo perturbation calculations of
    interaction energies
Phys. Rev. A **43**, 2139–2151 (1991)

82
X.-P. Li, D. M. Ceperley, & R. M. Martin
Cohesive energy of silicon by the Green's-function
    Monte Carlo method
Phys. Rev. B **44**, 10929–10932 (1991)

83
Z. Sun, R. N. Barnett, & W. A. Lester, Jr.
Optimization of a multideterminant wave function for quantum
    Monte Carlo: Li_2 (X $^1\Sigma_g^+$)
J. Chem. Phys. **96**, 2422–2423 (1992)

84
R. N. Barnett, P. J. Reynolds, & W. A. Lester, Jr.
Computation of transition dipole moments by Monte Carlo
J. Chem. Phys. **96**, 2141–2154 (1992)

85

J. B. Anderson

Quantum chemistry by random walk: Higher accuracy for H_3^+

J. Chem. Phys. **96**, 3702–3706 (1992)

86

V. Buch

Treatment of rigid bodies by diffusion Monte Carlo: Application to
the para-H_2—H_2O and ortho-H_2—H_2O clusters

J. Chem. Phys. **97**, 726–729 (1992)

87

D. L. Diedrich & J. B. Anderson

An accurate quantum Monte Carlo calculation of the barrier
height for the reaction $H + H_2 \rightarrow H_2 + H$

Science **258**, 786–788 (1992)

88

P. Ballone, C. J. Umrigar, & P. Delaly

Energies, densities, and pair correlation functions of jellium
spheres by the variational Monte Carlo method

Phys. Rev. B **45**, 6293–6296 (1992)

89

R. F. Bishop, E. Buendia, M. F. Flynn, & R. Guardiola

Diffusion Monte Carlo determination of the binding energy of
the ^{4}He nucleus for model Wigner potentials

J. Phys. G: Nucl. Part. Phys. **18**, L21–L27 (1992)

90

P. Belohorec, S. M. Rothstein, & J. Vrbik

Infinitesimal differential diffusion quantum Monte Carlo
study of CuH spectroscopic constants

J. Chem. Phys. **98**, 6401–6405 (1993)

91

D. M. Schrader, T. Yoshida, & K. Iguchi

Binding energies of positronium fluoride and positronium bromide
by the model potential quantum Monte Carlo method

J. Chem. Phys. **98**, 7185–7190 (1993)

92
J. B. Anderson, C. A. Traynor, & B. M. Boghosian
An exact quantum Monte Carlo calculation of
   the helium-helium intermolecular potential
J. Chem. Phys. **99**, 345–351 (1993)

93
A. Bhattacharya & J. B. Anderson
Exact quantum Monte Carlo calculation of the H-He interaction
   potential
Phys. Rev. A **49**, 2441–2445 (1994)

94
L. Mitáš
Quantum Monte Carlo calculation of the Fe atom
Phys. Rev. A **49**, 4411–4414 (1994)

95
A. B. Finnila, M. A. Gomez, C. Stenson, C. Sebenik, & J. D. Doll
Quantum annealing: A new method for minimizing
   multidimensional functions
Chem. Phys. Lett. **219**, 343–348 (1994)

96
M. Lewerenz & R. O. Watts
Quantum Monte Carlo simulation of molecular vibrations:
   Application to formaldehyde
Mol. Phys. **81**, 1075–1091 (1994)

97
L. Mitáš & R. M. Martin
Quantum Monte Carlo of nitrogen: Atom, dimer, atomic and
   molecular solids
Phys. Rev. Lett. **72**, 2438–2441 (1994)

98
F. Bolton
The effect of an uneven potential on the few-electron spectrum in
   a quantum dot
Physica B **212**, 218–223 (1995)

99
B. Chen & J. B. Anderson
Improved quantum Monte Carlo calculation of the ground-state
  energy of the hydrogen molecule
J. Chem. Phys. **102**, 2802–2805 (1995)

100
S. D. Kenny, G. Rajagopal, & R. J. Needs
Relativistic corrections to atomic energies from quantum
  Monte Carlo calculations
Phys. Rev. A **51**, 1898–1904 (1995)

101
C. Huiszoon & M. Caffarel
A quantum Monte Carlo perturbational study of
  the He-He interaction
J. Chem. Phys. **104**, 4621–4631 (1996)

102
R. E. Tuzun, D. W. Noid, & B. G. Sumpter
An internal coordinate quantum Monte Carlo method for
  calculating vibrational ground state energies and wave
  functions of large molecules: A quantum geometric
  statement function approach
J. Chem. Phys. **105**, 5494–5502 (1996)

103
B. Chen, M. A. Gomez, M. Sehl, J. D. Doll, & D. L. Freeman
Theoretical studies of the structure and dynamics of metal/hydrogen
  systems: Diffusion and path integral Monte Carlo
  investigations of nickel and palladium clusters
J. Chem. Phys. **105**, 9686–9694 (1996)

104
D. F. R. Brown, J. K. Gregory, & D. C. Clary
A method to calculate vibrational frequency shifts in
  heteroclusters: Application to N_2^+-He_n
J. Chem. Soc., Faraday Trans. **92**, 11–15 (1996)

105
S. D. Kenny, G. Rajagopal, R. J. Needs, W.-K. Leung,
M. J. Godfrey, A. J. Williamson, & W. M. C. Foulkes
Quantum Monte Carlo calculations of the energy of
the relativistic homogeneous electron gas
Phys. Rev. Lett. **77**, 1099–1102 (1996)

106
K. Liu, M. G. Brown, C. Carter, R. J. Saykally, J. K. Gregory, &
D. C. Clary
Characterization of a cage form of the water hexamer
Nature **381**, 501–503 (1996)

107
M. Lewerenz
Structure and energetics of small helium clusters: Quantum
simulations using a recent perturbational pair potential
J. Chem. Phys. **106**, 4596–4603 (1997)

108
R. N. Barnett, Z. Sun, & W. A. Lester, Jr.
Fixed-sample optimization in quantum Monte Carlo using
a probability density function
Chem. Phys. Lett. **273**, 321–328 (1997)

109
H.-J. Flad, M. Dolg, & A. Shukla
Spin-orbit coupling in variational quantum Monte Carlo
calculations
Phys. Rev. A **55**, 4183–4195 (1997)

110
R. Q. Hood, M. Y. Chou, A. J. Williamson, G. Rajagopal,
R. J. Needs, & W. M. C. Foulkes
Quantum Monte Carlo investigation of exchange and
correlation in silicon
Phys. Rev. Lett. **78**, 3350–3353 (1997)

111

V. N. Staroverov, P. Langfelder, & S. M. Rothstein

Monte Carlo study of core-valence separation schemes

J. Chem. Phys. **108**, 2873–2885 (1998)

112

D. Bressanini, M. Mella, & G. Morosi

Positronium chemistry by quantum Monte Carlo. I.
  Positronium-first row atom complexes

J. Chem. Phys. **108**, 4756–4760 (1998)

113

M. Zhao, D. Chekmarev, & S. A. Rice

Quantum Monte Carlo simulations of the structure in
  the liquid-vapor interface of BiGa binary alloys

J. Chem. Phys. **108**, 5055–5067 (1998)

114

T. Yoshida & G. Miyako

Quantum Monte Carlo with model potentials for molecules

J. Chem. Phys. **108**, 8059–8061 (1998)

115

C.-J. Huang, C. Filippi, & C. J. Umrigar

Spin contamination in quantum Monte Carlo wave functions

J. Chem. Phys. **108**, 8838–8847 (1998)

116

C. W. Greeff & W. A. Lester, Jr.

A soft Hartree-Fock pseudopotential for carbon with
  application to quantum Monte Carlo

J. Chem. Phys. **109**, 1607–1612 (1998)

117

F. Schautz, H.-J. Flad, & M. Dolg

Quantum Monte Carlo study of Be_2 and group 12 dimers
  M_2 (M = Zn, Cd, Hg)

Theor. Chem. Acc. **99**, 231–240 (1998)

118
H.-S. Lee, J. M. Herbert, & A. B. McCoy
Adiabatic diffusion Monte Carlo approaches for studies of ground
  and excited state properties of van der Waals complexes
J. Chem. Phys. **110**, 5481–5484 (1999)

119
Y. Shlyakhter, S. Sokolova, A. Lüchow, & J. B. Anderson
Energetics of carbon clusters C_8 and C_{10} from
  all-electron quantum Monte Carlo calculations
J. Chem. Phys. **110**, 10725–10729 (1999)

120
S. Baroni & S. Moroni
Reptation quantum Monte Carlo: A method for unbiased
  ground-state averages and imaginary-time correlations
Phys. Rev. Lett. **82**, 4745–4748 (1999)

121
M. W. Severson & V. Buch
Quantum Monte Carlo simulation of intermolecular excited
  vibrational states in the cage water hexamer
J. Chem. Phys. **111**, 10866–10874 (1999)

122
S. A. Alexander & R. L. Coldwell
Relativistic calculations using Monte Carlo methods:
  One-electron systems
Phys. Rev. E **60**, 3374–3379 (1999)

123
S. Broude, J. O. Jung, & R. B. Gerber
Combined diffusion quantum Monte Carlo–vibrational
  self-consistent field (DQMC-VSCF) method for
  excited vibrational states of large polyatomic systems
Chem. Phys. Lett. **299**, 437–442 (1999)

124
W. M. C. Foulkes, R. Q. Hood, & R. J. Needs
Symmetry constraints and variational principles in diffusion
  quantum Monte Carlo calculations of excited-state energies
Phys. Rev. B **60**, 4558–4570 (1999)

125
X. Lin, H. Zhang, & A. M. Rappe
Optimization of quantum Monte Carlo wave functions using
    analytical energy derivatives
J. Chem. Phys. **112**, 2650–2654 (2000)

126
J. B. Anderson
Quantum Monte Carlo: Direct calculation of corrections to trial
    wave functions and their energies
J. Chem. Phys. **112**, 9699–9702 (2000)

127
A. Sarsa, K. E. Schmidt, & J. W. Moskowitz
Constraint dynamics for quantum Monte Carlo calculations
J. Chem. Phys. **113**, 44–47 (2000)

128
C. Filippi & S. Fahy
Optimal orbitals from energy fluctuations in correlated wave
    functions
J. Chem. Phys. **112**, 3523–3531 (2000)

129
A. Lüchow & R. F. Fink
On the systematic improvement of fixed-node diffusion quantum
    Monte Carlo energies using pair natural orbital
    CI guide functions
J. Chem. Phys. **113**, 8457–8463 (2000)

130
J. C. Grossman, W. A. Lester, Jr., & S. G. Louie
Quantum Monte Carlo density functional theory characterization
    of 2-cyclopentenone and 3-cyclopentenone formation
    from $O(^3P)$ + cyclopentadiene
J. Am. Chem. Soc. **122**, 705–711 (2000)

131
F. Pederiva, C. J. Umrigar, & E. Lipparini
Diffusion Monte Carlo study of circular quantum dots
Phys. Rev. B **62**, 8120–8125 (2000)

132
D. C. Clary
Torsional diffusion Monte Carlo: A method for quantum
  simulations of proteins
J. Chem. Phys. **114**, 9725–9732 (2001)

133
S. Manten & A. Lüchow
On the accuracy of the fixed-node diffusion quantum Monte Carlo
  method
J. Chem. Phys. **115**, 5362–5366 (2001)

134
A. J. Williamson, R. Q. Hood, & J. C. Grossman
Linear-scaling quantum Monte Carlo calculations
Phys. Rev. Lett. **87**, 246406-1/4 (2001)

135
S. B. Healy, C. Filippi, P. Kratzer, E. Penev, & M. Scheffler
Role of electronic correlation in the Si(100) reconstruction:
  A quantum Monte Carlo study
Phys. Rev. Lett. **87**, 016105-1/4 (2001)

136
R. Baer
Ab initio computation of molecular singlet-triplet energy
  differences using auxiliary field Monte Carlo
Chem. Phys. Lett. **343**, 535–542 (2001)

137
J. C. Grossman
Benchmark quantum Monte Carlo calculations
J. Chem. Phys. **117**, 1434–1440 (2002)

138
F. Schautz & H.-J. Flad
Selective correlation scheme within diffusion quantum
  Monte Carlo
J. Chem. Phys. **116**, 7389–7399 (2002)

139
D. Blume
Fermionization of a bosonic gas under highly elongated
  confinement: A diffusion quantum Monte Carlo study
Phys. Rev. A **66**, 053613-1/7 (2002)

140
G. E. Astrakharchik & S. Giorgini
Quantum Monte Carlo study of the three- to one-dimensional
  crossover for a trapped Bose gas
Phys. Rev. A **66**, 053614-1/6 (2002)

141
K. E. Riley & J. B. Anderson
Higher accuracy quantum Monte Carlo calculations of the barrier
  for the $H + H_2$ reaction
J. Chem. Phys. **118**, 3437–3438 (2003)

142
S. Manten & A. Lüchow
Linear scaling for the local energy in quantum Monte Carlo
J. Chem. Phys. **119**, 13078–1312 (2003)

143
J. Carlson, J. Morales, Jr., V. R. Pandharipande, &
  D. E. Ravenhall
Quantum Monte Carlo calculations of neutron matter
Phys. Rev. C **68**, 025802-1/13 (2003)

144
C. A. Schuetz, M. Frenklach, A. C. Kollias, & W. A. Lester, Jr.
Geometry optimization in quantum Monte Carlo with
  solution mapping: Application to formaldehyde
J. Chem. Phys. **119**, 9386–9392 (2003)

145
M. Nekovee, W. M. C. Foulkes, & R. J. Needs
Quantum Monte Carlo studies of density functional theory
Math. Comput. Simulat. **62**, 463–470 (2003)

146
W. Schattke, R. Bahnsen, & R. Redmer
Variational quantum Monte-Carlo method in surface physics
Prog. Surf. Sci. **72**, 87–116 (2003)

147
A. Aspuru-Guzik, O. El Akramine, J. C. Grossman, &
   W. A. Lester, Jr.
Quantum Monte Carlo for electronic excitations of
   free-base porphyrin
J. Chem. Phys. **120**, 3049–3050 (2004)

148
P. Cazzato, S. Paolini, S. Moroni, & S. Baroni
Rotational dynamics of CO solvated in small He clusters:
   A quantum Monte Carlo study
J. Chem. Phys. **120**, 9071–9076 (2004)

149
F. Schautz, F. Buda, & C. Filippi
Excitations in photoactive molecules from
   quantum Monte Carlo
J. Chem. Phys. **121**, 5836–5844 (2004)

150
K. Hongo, R. Maezono, Y. Kawazoe, H. Yasuhara,
   M. D. Towler, & R. J. Needs
Interpretation of Hund's multiplicity rule for the carbon atom
J. Chem. Phys. **121**, 7144–7147 (2004)

151
D. L. Crittenden, K. C. Thompson, M. Chebib, &
   M. J. T. Jordan
Efficiency considerations in the construction of interpolated
   potential energy surfaces for the calculation of
   quantum observables by diffusion Monte Carlo
J. Chem. Phys. **121**, 9844–9854 (2004)

152
N. Goldman & R. J. Saykally
Elucidating the role of many-body forces in liquid water.
  I. Simulations of water clusters on the VRT(ASP-W)
  potential surfaces
J. Chem. Phys. **120**, 4777–4789 (2004)

153
S. Moroni, N. Blinov, & P.-N. Roy
Quantum Monte Carlo study of helium clusters
  doped with nitrous oxide: Quantum solvation and
  rotational dynamics
J. Chem. Phys. **121**, 3577–3581 (2004)

154
S.-I. Lu
The accuracy of diffusion quantum Monte Carlo simulations in
  the determination of molecular equilibrium structures
J. Chem. Phys. **121**, 10365–10369 (2004)

155
S. A. Alexander & R. L. Coldwell
A ground state potential energy surface for H_2 using Monte Carlo
methods
J. Chem. Phys. **121**, 11557–11561 (2004)

156
J. C. Grossman & L. Mitas
Efficient quantum Monte Carlo energies for molecular dynamics
  simulations
Phys. Rev. Lett. **94**, 056403-1/4 (2005)

157
S.-I. Lu
Theoretical study of transition state structure and reaction
  enthalpy of the $F + H_2 \to HF + H$ reaction
  by a diffusion quantum Monte Carlo approach
J. Chem. Phys. **122**, 194323-1/7 (2005)

158
A. Ghosal, C. J. Umrigar, H. Jiang, D. Ullmo, & H. U. Baranger
Interaction effects in the mesoscopic regime: A quantum
   Monte Carlo study of irregular quantum dots
Phys. Rev. B **71**, 241306-1/4 (2005)

159
A. Ma, M. D. Towler, N. D. Drummond, & R. J. Needs
Scheme for adding electron-nucleus cusps to Gaussian orbitals
J. Chem. Phys. **122**, 224322-1/7 (2005)

160
M. P. Nightingale & M. Moodley
Interdimensional degeneracies in van der Waals clusters
   and quantum Monte Carlo computation of rovibrational states
J. Chem. Phys. **123**, 014304-1/7 (2005)

161
Z. Xie, B. J. Braams, & J. M. Bowman
Ab initio global potential-energy surface for $H_5^+ \rightarrow H_3^+ + H_2$
J. Chem. Phys. **122**, 224307-1/9 (2005)

Quantum Monte Carlo

1

E. SCHRÖDINGER

Über die Umkehrung der Naturgesetze

Sitzber. Preuss. Akad. Wiss. Phys.-math. Kl., 144–153 (1931)[a]

A decade before his 1926 paper[b] reporting his discovery of wave mechanics, a derivation of the time-independent wave equation, and its solutions for hydrogen-like atoms, Erwin Schrödinger had carried out several studies dealing with Brownian motion. It is clear from several of the papers he wrote that he was completely familiar with diffusion equations, the Fokker-Planck equation in many forms, their connections with Green's functions, and with behavior of systems of particles undergoing random walks including those in force fields.[c] Thus, it is not surprising that Schrödinger was aware of the relation of the Schrödinger equation in imaginary time,

$$\frac{\partial \psi}{\partial \tau} = \frac{\hbar^2}{2m} \nabla^2 \psi - V\psi,$$

to the diffusion equation governing particles undergoing diffusion,

$$\frac{\partial C}{\partial t} = D\nabla^2 C.$$

In this paper, with the title "On the Reversal of Natural Laws," Schrödinger pointed out the remarkable formal analogy of diffusion with quantum mechanics and the discrepancies introduced by the factor $\sqrt{-1}$. In his discussion Schrödinger made the statement, "I am unable to foresee whether this analogy will prove itself useful in the clarification of quantum mechanical concepts," but he finished the paper with the hope it would. There is no indication that Schrödinger was aware of the possibilities of a next step into quantum Monte Carlo — the use of random walks to provide solutions to the wave equation. This was left for others to discover later.

[a]E. Schrödinger, *Collected Papers,* Volume 1, Austrian Academy of Sciences, Vienna, 1984, pp. 412–422.

[b]E. Schrödinger, Quantisierung als Eigenwertproblem, *Ann. Phys.* **79**, 361 (1926).

[c]W. Moore, *Schrödinger, Life and Thought,* Cambridge University Press, 1989, pp. 83, 99–102.

2

N. METROPOLIS & S. ULAM

The Monte Carlo method

J. Am. Stat. Assoc. **44**, 335–341 (1949)

In this paper Metropolis and Ulam gave a brief introduction to "the Monte Carlo method" which is described as a statistical approach to the study of differential equations as applied by Metropolis, Ulam, Fermi, von Neumann, Feynman, and others at the Los Alamos Laboratory in the 1940s.[a] Several examples of applications of Monte Carlo calculations are given. These include predicting the probability of winning at the game of solitaire, calculating the volume of an irregular region in high-dimensional space, and solving the Fokker-Planck equation for diffusion and multiplication of nuclear particles.

The paper is completed with the first published description of the *diffusion quantum Monte Carlo* method, which is attributed to a suggestion by Fermi. In their words, "... the time-independent Schrödinger equation

$$\nabla^2 \psi(x, y, z) = (E - V)\psi(x, y, z)$$

could be studied as follows. Re-introduce time dependence by considering

$$u(x, y, z, t) = \psi(x, y, z)e^{-Et}$$

u will obey the equation

$$\frac{\partial u}{\partial t} = \nabla^2 u - V u.$$

This last equation can be interpreted however as describing the behavior of a system of particles each of which performs a random walk, i.e., diffuses isotropically and at the same time is subject to multiplication, which is determined by the point value of V." The authors then indicate that the procedure gives the wavefunction corresponding to the lowest eigenvalue E.

---

[a] Similar summaries by the authors are: N. Metropolis, in *Proceedings, Symposium on Monte Carlo Methods* (H. A. Meyer, ed.), Wiley, New York, 1956, pp. 29–36; S. M. Ulam, *A Collection of Mathematical Problems*, Interscience, New York, 1960, pp. 123–128.

3

M. H. KALOS

Monte Carlo calculations of the ground state
of three- and four-body nuclei

Phys. Rev. **128**, 1791–1795 (1962)

In this paper Kalos proposed what is now known as the *Green's function quantum Monte Carlo* (GFQMC) method for solving the Schrödinger equation. Kalos made the "critical" observation that the time-independent Schrödinger equation could be transformed into an integral equation for which the Green's function is known if the boundary conditions correspond to the wavefunction vanishing at infinity. In a $3n$-dimensional space R the integral form is (modified here)

$$\psi(R) = \int G(R', R) \frac{V(R')}{E} \psi(R') dR',$$

an equation which may be simulated by a random walk process in which particles at positions R' are altered in weight by the V/E term and moved to new positions R obtained by sampling the Green's function $G(R', R)$. As for simple diffusion, repetition leads to a distribution of particles corresponding to the lowest-energy wavefunction from which the energy E may be determined. Since time is not involved, there is no time-step error associated with the procedure. The term V/E must remain positive if mixed signs for particle weights are to be avoided.[a] Kalos examined the utility of the method in determining the boson ground-state energies for several three- and four-body nuclear problems involving pairwise-additive interactions with square, Gaussian, and exponential wells. The errors in the computed binding energies were estimated to be about 1%.

[a] Several other devices for avoiding the problem of mixed signs in small systems are available. See J. B. Anderson, C. A. Traynor, and B. M. Boghosian, Quantum chemistry by random walk: Exact treatment of many-electron systems, *J. Chem. Phys.* **95**, 7418 (1991).

4

H. CONROY

Molecular Schrödinger equation.
II. Monte Carlo evaluation of integrals

J. Chem. Phys. **41**, 1331–1335 (1964)

This paper and its three companion papers[a] published back-to-back in the *Journal of Chemical Physics* describe the first variational QMC calculations for molecular systems. The first paper introduces a general form for a one-electron wavefunction along with discussions of the requirements for an accurate wavefunction and the procedure for optimization by minimizing the variance in local energies. The second, with the title given above, describes the Monte Carlo evaluation of the matrix elements required for determination of the expectation value of the energy in a variational calculation. As pointed out, Monte Carlo schemes had often been used for integrations, but there was apparently no prior report of such schemes for problems in quantum mechanics. The need for selecting points from an approximate ψ^2 distribution was recognized and a procedure for doing so was proposed. In this paper and in the third and fourth papers the methods were illustrated with applications for the systems H_2^+, HeH^{2+}, HeH^+, H_3^{2+}, and H_3^+. In the case of H_3^+ sections of potential energy surfaces were determined for linear and triangular nuclear configurations using wavefunctions with up to 29 terms. The minimum-energy structure for H_3^+ was found to be an equilateral triangle of side length 1.68 bohr with an energy of -1.357 hartrees, values more accurate than those of any prior calculations.

[a] H. Conroy, Molecular Schrödinger equation. I to IV, *J. Chem. Phys.* **41**, 1327, 1331, 1336, 1341 (1964).

5

W. L. McMILLAN

Ground state of liquid ^{4}He

Phys. Rev. **138**, A442–A451 (1965)

One of the earliest uses of Metropolis sampling[a] in variational QMC calculations was not in an electronic structure problem, but in the calculation of the energy of a sample of liquid helium in which the atoms were treated as single particles. The interaction among the atoms was specified as a pairwise-additive Lennard-Jones 6–12 potential, which is a rather good approximation in this case. The trial wavefunction was a product form of the type $\prod_{i<j} f(r_{ij})$, with $f(r) = \exp\left[-(a_1/r)^{a_2}\right]$ and variable parameters a_1 and a_2. The ground-state wavefunction for this collection of bosons is nodeless. Calculations were carried out with 32 to 108 atoms at several lattice densities in a cube with periodic boundary conditions. The energies so determined were within about 20% of the experimental values and two-body correlation functions were in reasonable agreement with those derived from x-ray data. This first variational QMC calculation for liquid helium was followed within a few years by both Green's function and diffusion QMC calculations.

[a]N. Metropolis, A. W. Rosenbluth, M. N. Rosenbluth, A. H. Teller, and E. Teller, Equation of state calculations by fast computing machines, *J. Chem. Phys.* **21**, 1087 (1953).

6

M. H. KALOS

Stochastic wave function for atomic helium

J. Comput. Phys. **1**, 257–276 (1967)

This paper reports a first application of the Green's function approach (GFQMC) introduced by Kalos[a] to the calculation of the wavefunction for an atomic system. Although the precision of the results obtained was (far) "less than that already achieved for helium in a number of existing calculations," the study provided a useful exploration of the method. It revealed some of the problems associated with the sign change for particle weights introduced by the potential energy term (V/E) occurring in the expression for sampling with a Green's function. Kalos investigated several methods for avoiding or minimizing the problem. Since the ground-state wavefunction for helium is nodeless, the sign problem associated with fermion nodes does not present itself in this case.

[a]M. H. Kalos, Monte Carlo calculations of the ground state of three- and four-body nuclei, *Phys. Rev.* **128**, 1791 (1962).

7

R. C. GRIMM & R. G. STORER

Monte-Carlo solution of Schrödinger's equation

J. Comput. Phys. **7**, 134–156 (1971)

With this paper Grimm and Storer introduced importance sampling to the field of quantum Monte Carlo. They had earlier used numerical quadrature for estimating the integrals required in evaluating energies in a diffusion-like path integral iteration scheme for solving the Schrödinger equation.[a] This was extended to larger numbers of particles and higher dimensions with use of Monte Carlo methods. In conjunction with this they recognized they could take advantage of prior knowledge of the general form of a wavefunction to obtain more accurate estimates of the eigenvalue E_0. With an importance sampling function $\psi_T(X)$ approximating the true wavefunction $\psi(X)$, the eigenvalue is given by

$$E_0 = \frac{\int \psi(X)\psi_T(X)\frac{H\psi_T(X)}{\psi_T(X)}dX}{\int \psi(X)\psi_T(X)dX}.$$

For particle distributions corresponding to ψ generated by a quantum Monte Carlo procedure the eigenvalue is estimated as

$$E_0 \approx \frac{\sum \psi_T(X)\frac{H\psi_T(X)}{\psi_T(X)}}{\sum \psi_T(X)}.$$

where the sums are for equal-weight particles in the ψ distribution. The authors noted that there would be zero variance in the estimate of E_0 with ψ_T actually equal to ψ and a very small variance with ψ_T a good approximation to ψ.

Also included is a description of revisons to the sampling procedure to produce a set of particles representing the $\psi(X)\psi_T(X)$ distribution.

[a]R. Grimm and R. G. Storer, A new method for the numerical solution of the Schrödinger equation, *J. Comput. Phys.* **4**, 230 (1969).

8

M. H. KALOS, D. LEVESQUE, & L. VERLET

Helium at zero temperature with hard-sphere and other forces

Phys. Rev. A **9**, 2178–2195 (1974)

In using the Green's function QMC method for calculations of hard-sphere systems one encounters the problem of multiplying the weight of a walker by the term V/E or (potential energy)/(energy) when two spheres overlap and the potential energy is infinity. The authors of this paper note that this is "the only — but very great — complication in the Green's function," and they were able to devise a sampling method using subdomains to overcome the problem. Calculations were carried out for 256 particles in three-dimensional boxes of several sizes. Periodic boundary conditions were imposed by replacing exiting walkers at the opposite sides of the boxes. Importance sampling was used to aid in controlling the weights of walkers, as with a drift term in diffusion QMC, and in evaluating the energies. The trial function for importance sampling was a simple product of terms for all particle-particle distances. Both fluids and fcc crystals were treated at several particle densities. The Green's function calculations produced energies slightly but significantly lower than those of corresponding variational QMC calculations.

The treatment for hard spheres was extended to Lennard-Jones 6–12 particles simulating helium atoms with use of a perturbation approach. The energies calculated in this way gave reasonable agreement with experimental values except at high densities.

9

K. S. LIU, M. H. KALOS, & G. V. CHESTER

Quantum hard spheres in a channel

Phys. Rev. A **10**, 303–308 (1974)

One of the early applications of Green's function QMC is that described in this paper for systems of hard spheres confined in a channel. The GFQMC method was applied to a collection of up to 64 hard spheres in a rectangular box, with rigid repulsive walls at top and bottom and periodic boundary conditions at the sides. Earlier variational QMC studies[a] provided good trial functions for importance sampling and variational upper bounds to the energies. With importance sampling the density of walkers corresponds to a population proportional to the product $\psi\psi_T$ of the exact wavefunction ψ and the trial wavefunction ψ_T. This is especially useful for computing the energy, but for a particle density (of hard spheres in this case) the square ψ^2 of the wavefunction is desired. The authors provide a clever solution to the problem in showing that in successive applications of Green's function sampling the weight of descendents of a walker becomes proportional to ψ/ψ_T. Thus, if a set of walkers representing $\psi\psi_T$ has its weights multiplied by the factor ψ/ψ_T, the new distribution represents ψ^2. With this procedure nearly exact hard-sphere distributions were obtained for a number of channel widths. The most striking feature observed was a layered structure across the channel much more pronounced than that predicted by the (inexact) variational calculations.

[a] K. S. Liu, M. H. Kalos, and G. V. Chester, Hard sphere model of the helium film, *J. Low Temp. Phys.* **13**, 227 (1973).

10

J. B. ANDERSON

A random-walk simulation of the Schrödinger equation:
H_3^+

J. Chem. Phys. **63**, 1499–1503 (1975)

This paper opened the field of electronic structure calculations
to quantum Monte Carlo calculations of the form originally suggested by Fermi.[a] The system treated was the molecular ion H_3^+
which has served as a test case for new methods in quantum mechanics since about 1935. With the three protons fixed in position
the problem is reduced to that of two electrons in three dimensions
each. For the ground state the electrons (fermions) have opposite spins and the wavefunction is nodeless. In atomic units the
Schrödinger equation in imaginary time becomes

$$\frac{\partial \psi}{\partial \tau} = \frac{1}{2}\nabla_1^2 \psi + \frac{1}{2}\nabla_2^2 \psi - (V - V_{ref})\psi \overset{ss}{=} -(E - V_{ref})\psi.$$

The random walk of particles was executed in six dimensions with
step sizes selected from a Gaussian distribution appropriate for a
finite time step $\Delta\tau$, and the multiplication term was simulated by
a probabilistic birth-death process. The overall number of particles
was controlled in part by the choice of the reference potential energy V_{ref}. The energy was determined from the growth rate under
conditions of steady-state shape of the distribution as suggested
by the second equality $\overset{ss}{=}$. Extrapolation of energies to that for a
time-step size of zero was required.

The energy obtained for the equilibrium equilateral triangle nuclear configuration was lower (but with an overlapping error bar)
than the lowest-energy analytic variational result at the time and
in agreement with recent predictions. The author discussed several
of the questions bearing on the utility of the method, proposed the
fixed-node method for incorporating the Pauli exclusion principle,
and concluded that "'In general, not enough is known for confi-
dent assessment of the merits of the method, but initial results are
encouraging."

[a]N. Metropolis and S. Ulam, The Monte Carlo method, *J. Am. Stat. Assoc.* **44**, 335 (1949).

11

D. J. KLEIN & H. M. PICKETT

Nodal hypersurfaces and Anderson's random-walk
simulation of the Schrödinger equation

J. Chem. Phys. **64**, 4811–4812 (1976)

Klein and Pickett provided a discussion of the problem of determining the nodes for a fixed-node QMC calculation proposed earlier
by Anderson.[a] They suggested using the nodes of a Hartree-Fock
wavefunction and noted that a Jastrow wavefunction[b] consisting of
a Hartree-Fock function multiplied by a permutationally symmetric and positive function to correct for electron correlation could be
quite accurate. Only later was it realized that the node structures
given by Hartree-Fock wavefunctions were surprisingly simple even
for large many-electron systems. Combinations of determinantal
functions with Jastrow functions were found especially useful as
trial functions for importance sampling in diffusion and Green's
function QMC calculations.

[a] J. B. Anderson, A random-walk simulation of the Schrödinger equation:
H_3^+, *J. Chem. Phys.* **63**, 1499 (1975).

[b] R. Jastrow, Many-body problem with strong forces, *Phys. Rev.* **98**, 1479
(1955). These functions were originally suggested by Bijl, but they are now
universally called "Jastrow functions." A. Bijl, The lowest wave function of
the symmetrical many particles system, *Physica* **7**, 869 (1940). More recently
the term "Jastrow function" has most often been used for the correlation term
itself without the Hartree-Fock function.

12

J. B. ANDERSON

Quantum chemistry by random walk. H ^2P, H$_3^+$ D_{3h} $^1A_1'$, H$_2$ $^3\Sigma_u^+$, H$_4$ $^1\Sigma_g^+$, Be ^1S

J. Chem. Phys. **65**, 4121–4127 (1976)

Applications of a *fixed-node method* for treating the node problem for excited states and for many-electron ground states were explored in this study. For cases of several particles in wells of one dimension and for the helium atom, the excited atom H ^2P, the excited molecule H$_2$ $^3\Sigma_u^+$, the many-electron system linear H$_4$ $^1\Sigma_g^+$, and the atom Be ^1S, it was shown that the antisymmetry requirements of the Pauli principle could be satisfied by the imposition of boundary conditions on the wavefunction. In the case of H$_2$ $^3\Sigma_u^+$, the six-dimensional electron configuration space is divided into two regions, one positive and one negative, separated by a single nodal surface. In a fixed-node diffusion calculation these nodes or boundaries act as sinks for walkers. The calculations are variational in nature, yielding (in the absence of any other errors) the exact energy for an exact specification of the nodal surface, and an energy higher than exact for an inexact nodal surface. For the several systems examined the energies so calculated were found to be within their statistical errors of the best estimates of the energies for these systems.

The paper also reported several extensions to reduce the error associated with finite time-steps and with possible node crossing and recrossing.

13

R. L. COLDWELL & R. E. LOWTHER

Monte Carlo calculation of the Born-Oppenheimer
potential between two helium atoms using
Hylleraas-type electronic wave functions

Int. J. Quantum Chem., Symp. Ser. **12**, 329–341 (1978)

Solution of the Schrödinger equation using the Rayleigh-Ritz variational method in combination with numerical integration was introduced as early as 1942 by Frost.[a] Conroy[b] added the Monte Carlo aspect with random selection of points in the configuration space of electrons for molecular systems. This paper and a related paper[c] by the same authors reported calculations similar to those of Conroy using variational QMC to determine the electronic structure of the helium dimer. Along with this Coldwell and Lowther introduced the idea of correlated sampling for such Monte Carlo variational calculations. They used a "biased selection" procedure to select 6400 points from an approximate ψ^2 distribution for evaluation of local energies and calculation of the expectation value of the energy for four internuclear distances. The trial functions were antisymmetrized products of two Hylleraas atomic-helium functions in combination with an exponential term containing dipole-dipole and dipole-quadrupole interaction terms. The energies obtained with reasonably good trial functions gave expectation values of the energy well below the Hartree-Fock limit and in agreement within their fairly large uncertainties with experimentally based estimates of the interaction energies for two helium atoms.

[a] A. A. Frost, The approximate solution of Schrödinger equations by a least squares method, *J. Chem. Phys.* **10**, 240 (1942).

[b] H. Conroy, Molecular Schrödinger equation. II. Monte Carlo evaluation of integrals, *J. Chem. Phys.* **41**, 1331 (1964).

[c] R. E. Lowther and R. L. Coldwell, Monte-Carlo calculation of the Born-Oppenheimer potential between two helium atoms, *Phys. Rev. A* **22**, 14 (1980).

14

J. B. ANDERSON

Quantum chemistry by random walk: H_4 square

Int. J. Quantum Chem. **15**, 109–120 (1979)

Early fixed-node QMC calculations for the H_4 system helped resolve an apparent disagreement between experimental measurements and theoretical predictions for the exchange reaction of H_2 and D_2. A number of rate measurements for the reaction $H_2 + D_2 \rightarrow 2\,HD$ indicated an activation energy of 35 to 45 kcal/mole and suggested a direct bimolecular reaction mechanism,[a] but several analytic variational calculations indicated barrier heights of greater than 115 kcal/mole for a variety of reaction paths. This paper reported fixed-node diffusion QMC calculations for the case of the H_4 square of side length 2.4 bohr in the $^1B_{1g}$ and $^1B_{2g}$ states. Node structures were taken from variational calculations for single- and multiple-determinant wavefunctions in the simplest forms meeting the symmetry requirements for these states. For the $^1B_{1g}$ state the variational calculation yielded a barrier of 186 kcal/mole relative to the exact value for separated H_2 and D_2. The QMC result was lower with a barrier of 162 ± 16 kcal/mole. A second calculation with a greatly simplified node structure gave 176 ± 14 kcal/mole for the barrier height. There was no suggestion of a reaction barrier less than 120 kcal/mol. For the $^1B_{2g}$ state the fixed-node QMC energy was approximately 65 kcal/mol lower than the expectation value for the trial wavefunction. The QMC calculations confirmed the earlier theoretical estimates of the barrier height and the experimental results were subsequently reinterpreted.

[a]S. H. Bauer and E. Ossa, Isotope exchange rates. III. The homogeneous four-center reaction $H_2 + D_2$, *J. Chem. Phys.* **45**, 434 (1966).

15

J. B. ANDERSON & B. H. FREIHAUT

Quantum chemistry by random walk: Method of successive corrections

J. Comput. Phys. **31**, 425–437 (1979)

This work extended the simple diffusion QMC method of solving the Schrödinger equation for a wavefunction ψ to that of solving for a correction δ to a trial wavefunction ψ_0 such that $\psi = \psi_0 + \delta$. The method may be applied in the manner of successive corrections to obtain very high accuracies. As described, the method is limited to solutions for systems with fixed nodes.

In the imaginary-time Schrödinger equation $\partial\psi/\partial\tau = \frac{\hbar^2}{2m}\nabla^2\psi - V\psi$, the term ψ may be replaced by $\psi_0 + \delta$ to obtain a similar equation for δ, $\partial\delta/\partial\tau = \frac{\hbar^2}{2m}\nabla^2\delta - V\delta + [\frac{\hbar^2}{2m}\nabla^2\psi_0 - V\psi_0]$, in which the last group corresponds to a distributed feed of positive and negative walkers. With cancellation of positive and negative walkers based on close proximity, or based on equivalent distributions for walkers having a long time to equilibrate after being fed, a steady-state distribution corresponding to the correction δ may be obtained. The energy may be determined from the growth rate at steady state. If it is possible to fit an analytic expression to the walker distribution, the original trial function ψ_0 can be revised to obtain a new trial function $\psi_1 = \psi_0 + \delta_0$, and successive corrections δ_1, δ_2, δ_3, $\cdots$ may be obtained. The feed term, the correction term, and the uncertainty in the energy become small with increasing accuracy in the trial function.

The method was illustrated for the cases of the particle-in-a-box and the hydrogen atom. For the latter, successive corrections improved the accuracy in energy by about a factor of ten with each iteration. For higher dimensional systems, difficulties in the fitting procedure may be expected to limit the applicability of the method.

16

Y. TOMASHIMA & J. OZAKI

Monte Carlo solution of Schrödinger's equation for the
hydrogen atom in a magnetic field

J. Comput. Phys. **33**, 382–396 (1979)

The problem of the hydrogen atom in a magnetic field can be
treated satisfactorily by approximate methods which are quite accurate for very low or very high magnetic field strengths. Before this paper the only useful treatments for intermediate field
strengths were variational methods giving upper bounds to the
energies. As described in the paper the diffusion QMC method
was applied in the same way as in earlier solutions for the hydrogen atom, but with additional magnetic terms added to the
Hamiltonian expression. These include complex expressions and
the required wavefunction includes a corresponding factor. With
the magnetic field aligned along one axis, integration over the angle
ϕ in (r, θ, ϕ) coordinates reduces the additional terms to real functions of r and z as well as the magnetic quantum number m. In
the diffusion/multiplication processes of a random walk procedure
these may be simply included with the potential energy term. In
this study the energy at each step was determined from the average multiplication factor for the walkers. Importance sampling was
not used. Results were obtained for quantum numbers $m = 0, 1$
at each of three magnetic field strengths. A comparison of results
showed good agreement with the lowest values obtained in several
of the earlier analytic variational calculations.

17

J. B. ANDERSON

Quantum chemistry by random walk: Higher accuracy

J. Chem. Phys. **73**, 3897–3899 (1980)

In an earlier paper[a] a method for the direct calculation of the difference between a true wavefunction and a trial wavefunction was described for simple diffusion calculations. This paper reports further increases in accuracy obtained by a combination of this difference method with importance sampling. The combination allows the calculation of corrections to the product $\psi\psi_T$ of true and trial wavefunctions, rather than to the wavefunction ψ itself. Using a new function $g = \psi\psi_T - \psi_T^2$, the original diffusion equation with drift and local energies is altered by replacement of $\psi\psi_T$ with g and by addition of a distributed source term. Since walkers fed by sampling the source term may be positive or negative, cancellation is required, but both sets of walkers proceed to the same distribution and age-based cancellation may be used. Node locations are not changed and remain at their original locations as specified by the trial function. The corrected wavefunction and the energy obtained are the same as for a fixed-node diffusion calculation using the same trial function for importance sampling. The method is useful in reducing statistical error, and because the statistical error occurs only in the difference terms, it is most effective when the trial wavefunction is already reasonably accurate. In test calculations for the hydrogen atom and the helium atom, both nodeless, the statistical error in the QMC calculations was greatly reduced. Thus, for hydrogen a trial function with an expectation value of -0.4998 hartrees led to a QMC energy of $-0.499995(4)$ hartrees.

[a] J. B. Anderson and B. H. Freihaut, Quantum chemistry by random walk: Method of successive corrections, *J. Comput. Phys.* **31**, 425 (1979).

18

D. M. CEPERLEY & B. J. ALDER

Ground state of the electron gas by a stochastic method

Phys. Rev. Lett. **45**, 566–569 (1980)

A simple way of incorporating sampling into diffusion quantum Monte Carlo calculations was introduced with this paper. Coupled with a nodal relaxation scheme it provided a means for calculating the ground-state energies of a collection of electrons (fermions) over a range of densities. The simulation method may be regarded as a more complex version of the simple diffusion method[a] or a less complex version of the Green's function approach.[b] With use of a trial function ψ_T the diffusion equation may be written in terms of a walker or particle density based on $f = \psi\psi_T$, the product of the true wavefunction ψ and the trial wavefunction ψ_T. The resulting equation includes a diffusion, a drift, and a multiplication term to take advantage of local energies as described by Grimm and Storer.[c] The calculations were carried out at first with fixed nodes to establish an upper bound to the exact fermion energy and starting positions for walkers in a released-node calculation. Walkers crossing a node were continued with a sign change to produce an overall distribution proceeding toward the boson ground state but providing at intermediate times an approximation to the fermion ground state. If relaxation to the fermion ground state is fast compared to further relaxation to the boson ground state, the fermion ground-state energy can be determined. For the electron gas at low densities this is the case and useful energies may be obtained. For this paper energies were calculated for a range of densities with systems of 38 to 246 particles and extrapolated to provide results for systems of infinite size.

[a] J. B. Anderson, A random-walk simulation of the Schrödinger equation: H_3^+, *J. Chem. Phys.* **63**, 1499 (1975).

[b] M. H. Kalos, Stochastic wave function for atomic helium, *J. Comput. Phys.* **1**, 257 (1967).

[c] R. C. Grimm and R. G. Storer, Monte-Carlo solution of Schrödinger's equation, *J. Comput. Phys.* **7**, 134 (1971).

19

F. MENTCH & J. B. ANDERSON

Quantum chemistry by random walk: Importance sampling for H_3^+

J. Chem. Phys. **74**, 6307–6311 (1981)

In this paper the combination of importance sampling[a] with diffusion QMC was shown to be remarkably effective in improving the accuracy of electonic structure calculations for molecular systems. The key to success was the use of a simple, fairly accurate trial wavefunction ψ_T for which the function itself, its derivatives and the local energy E_{loc} may be easily computed. This, used in combination with the diffusion-with-drift expression for walkers corresponding to the product $f = \psi\psi_T$ according to

$$\frac{\partial f}{\partial \tau} = \frac{\hbar^2}{2m}[\nabla^2 f - 2\nabla(f\nabla\ln\psi_T)] - [\frac{H\psi_T}{\psi_T} - E_{ref}]f$$

allows the energy to be evaluated as an average for all walkers of every step after reaching a fluctuating steady state. Since there is a time-step error associated with the simulation of diffusion, drift, and multiplication, varying the time-step and an extrapolation to zero step size are required. The result for H_3^+ in its triangular configuration of side length 1.6500 bohr was -1.3439 ± 0.0002 hartrees, a value well below the lowest variational energy levels at the time and having an error bar straddling the best modern value of $-1.343\,835$ hartrees. The use of importance sampling along with higher computer speeds resulted in an uncertainty about a factor of 60 lower than that of the previous calculation.[b]

---

[a]R. C. Grimm and R. G. Storer, Monte-Carlo solution of Schrödinger's equation, *J. Comput. Phys.* **7**, 134 (1971).

[b]J. B. Anderson, Quantum chemistry by random walk. H ^2P, H_3^+ D_{3h} $^1A_1'$, H_2 $^3\Sigma_u^+$, H_4 $^1\Sigma_g^+$, Be ^1S , *J. Chem. Phys.* **65**, 4121 (1976).

20, 21

K. McDOWELL & J. D. DOLL

Quantum Monte Carlo and the Hydride Ion

Chem. Phys. Lett. **82**, 127–129 (1981)

K. McDOWELL

Assessing the Quality of a Wavefunction using Quantum Monte Carlo

Int. J. Quantum Chem., Symp. Ser. **15**, 177–181 (1981)

These two papers introduced the term "Quantum Monte Carlo" to the scientific literature. Earlier in the same year McDowell[a] and Berne[b] had used the term in abstracts for an American Chemical Society meeting, and later in the year Gordon, Rothstein, and Proctor[c] used it in the title of a paper submitted to the *Journal of Computational Physics*.

The papers report investigations of variational QMC methods with Metropolis sampling using the hydride ion H^- and the helium atom He as examples. As had been found by earlier workers it is shown that relatively simple trial functions explicitly incorporating the interelectron distance r_{ij} can produce a high level of accuracy comparable to those of configuration interaction (CI) calculations based on many-term trial functions without r_{ij}.

[a]K. McDowell, Application of Quantum Monte Carlo to Molecular Systems, *Abstracts of Papers of the American Chemical Society* **181**, 30-Phys (1981).

[b]B. J. Berne, Quantum Monte Carlo and Molecular Dynamic Studies of Polyatomic Molecules in Liquids, *Abstracts of Papers of the American Chemical Society* **181**, 87-Phys (1981).

[c]H. L. Gordon, S. M. Rothstein, and T. R. Proctor, Efficient Variance Reduction Transformations for the Simulation of a Ratio of Two Means – Application to Quantum Monte Carlo Simulations, *J. Computational Phys.* **47**, 375 (1982).

22

J. G. ZABOLITZKY & M. H. KALOS

Solution of the four-nucleon Schrödinger equation

Nuc. Phys. A **356**, 114–128 (1981)

An application of the Green's function QMC method to the problem of four nucleons is reported in this paper. The calculations were carried out for the case of four nucleons interacting according to a pairwise-additive potential given by the sum of the Malfliet-Tjon interaction terms for each pair. The problem was reduced to that of four interacting, spinless bosons and the ground state is nodeless. The calculation was a straightforward application of Green's function QMC as described by Kalos[a] with avoidance of sign changes for walkers by a large shift in the zero of potential energy. Energies were evaluated from the growth rate of walkers and with use of importance sampling using a function of the Jastrow form. The energy determined for the case of the α-particle was -31.1 ± 0.2 MeV for a trial function with an expectation value of -28.8 ± 0.01 MeV.

Additional calculations were performed for the case of the four-atom helium droplet or cluster He_4 with the atoms interacting via pairwise-additive Lennard-Jones potentials. The energy for the cluster was found to be -0.39 ± 0.01 K, a value in agreement with earlier estimates for similar Lennard-Jones potentials. The authors pointed out that the results of QMC calculations such as these could serve as benchmarks for less rigorous methods.

[a]M. H. Kalos, Energy of a boson fluid with Lennard-Jones potentials, *Phys. Rev. A* **2**, 250 (1970).

23

D. M. ARNOW, M. H. KALOS, M. A. LEE, & K. E. SCHMIDT

Green's function Monte Carlo for few fermion problems

J. Chem. Phys. **77**, 5562–5572 (1982)

A method for treating the nodes for small systems was described in this paper. The general procedure involves the use of both positive and negative walkers in Green's function QMC, along with cancellation of walkers of opposite sign in close proximity. The method is illustrated for the first excited state of a particle in a one-dimensional box and for three model problems with three particles in three dimensions, interacting with each other through square-well potentials with repulsive and attractive parts.

The calculations were executed with the Green's function procedure, made convenient by availability of known Green's functions satisfying the conditions of the sample problems. A partial cancellation of positive and negative walkers was made on the basis of overlapping Green's functions, and the distributions of steps for the walkers involved were altered accordingly. Without cancellations, the distributions of positive and negative walkers proceed to the boson ground-state distribution. With cancellation and with control of their numbers, the positive and negative walkers tend to accumulate in separate regions and to commingle in the vicinity of a node. The arithmetic sum of the walkers yields the excited or fermion wavefunction with a node separating positive and negative regions. For the model problems the method successfully produced accuracies within a few percent of the exact energies, where known. It was noted that for larger problems, the total number of walkers required to maintain an adequate density for sufficient cancellation might be prohibitive.

24

J. W. MOSKOWITZ, K. E. SCHMIDT, M. A. LEE, & M. H. KALOS

Monte Carlo variational study of Be: A survey of correlated wavefunctions

J. Chem. Phys. **76**, 1064–1067 (1982)

This paper reports a successful attempt to construct simple but accurate wavefunctions for an atomic system by taking advantage of explicit electron-electron distances r_{ij} which can be incorporated into wavefunctions intended for QMC calculations which do not require analytic integrations. The principal advance reported is success with the use of functional forms more complex than that of earlier "Jastrow" functions such as $\exp[br_{ij}/(1+cr_{ij})]$. In this case the forms are the "transcorrelated" expressions of Boys and Handy[a] which include exponential terms involving $r_{ij}^m r_{ik}^n$, mixed powers of electron-electron and electron-nucleus distances. The results were not particularly impressive because the coefficients within these terms were not optimized. Multiplying a Slater determinant by a typical Jastrow term tends to favor the overall expansion of the electron cloud as it keeps electrons apart, but it usually fails to give much improvement unless a corresponding adjustment shrinking the cloud is made. The slight improvements for the Be atom seen in this work were soon superceded in the authors' later work in which such adjustments produced very much improved wavefunctions.[b]

[a]S. F. Boys and N. C. Handy, A condition to remove indeterminacy in interelectronic correlation functions, *Proc. Roy. Soc. London, Ser. A* **309**, 209 (1969).

[b]K. E. Schmidt, J. W. Moskowitz, Correlated Monte-Carlo wave-functions for the atoms He through Ne, *J. Chem. Phys.* **93**, 4172 (1990).

25

J. W. MOSKOWITZ, K. E. SCHMIDT, M. A. LEE, & M. H. KALOS

A new look at correlation energy in atomic and molecular systems. II. The application of the Green's function Monte Carlo method to LiH

J. Chem. Phys. **77**, 349–355 (1982)

This paper describes the use of a combination of importance sampling with fixed-node diffusion QMC for a molecular system. Although the two devices had been used separately for several four-electron systems,[a] the combination gives the important advantages of low values of multiplication of walkers and nearly complete elimination of node crossing. The authors derive the Green's function approach to importance sampling and use the short-time approximation to obtain a calculation procedure essentially the same as that for diffusion with drift. The calculations for LiH were carried out with two different trial functions ψ_T specifying the node locations: an SCF function of Slater orbitals and a generalized valence bond function, each multiplied by a simple Jastrow function. This allowed a comparison of results for the two node structures, accompanied with the recognition that the energies represented upper bounds to the true energy which would be given for the correct node locations. At the equilibrium internuclear distance, the calculated energies were the same within their statistical uncertainties. A complete potential energy curve was plotted for the generalized valence bond case. At the equilibrium distance of 3.015 bohr, the energy found was -8.0727 ± 0.0006 hartrees, with an additional uncertainty of about ± 0.01 hartree due to time-step error. Within the uncertainties this calculated energy agrees with more recent estimates[b] of the exact Born-Oppenheimer energy at 3.015 bohr.

[a] J. B. Anderson, Quantum chemistry by random walk: H_4 square, *Int. J. Quantum Chem.* **15**, 109 (1979).

[b] W. Cencek and J. Rychlewski, Benchmark calculations for He_2^+ and LiH molecules using explicitly correlated Gaussian functions, *Chem. Phys. Lett.* **320**, 549 (2000).

26

P. J. REYNOLDS, D. M. CEPERLEY, B. J. ALDER, & W. A. LESTER, JR.

Fixed-node quantum Monte Carlo for molecules

J. Chem. Phys. **77**, 5593–5603 (1982)

This paper extends earlier calculations for the molecules H_2 and LiH, and it increases the range to six and ten electrons with calculations for Li_2 and H_2O. The calculations were carried out with importance sampling using trial functions consisting of Slater determinants multiplied by Jastrow factors. For each system the expectation values of the energies for the trial functions were somewhat higher than those of the best configuration interaction calculations available at the time, but the fixed-node diffusion energies were significantly lower. The observed energy obtained for H_2 was -1.174 ± 0.001 hartrees, compared to earlier values of -1.1744 hartrees in analytic variational calculations. For H_2O it was -76.377 ± 0.007 hartrees, compared to a value of -76.3683 hartrees from variational calculations and an estimated true value for H_2O of -76.438 hartrees. The difference between the QMC value and the true value illustrates the problem of inaccurate node locations, in this case corresponding to a node location error of about 38 kcal/mol. Nevertheless, the energy so determined was about 5 kcal/mol lower than that of the lowest energy analytic variational calculation of the time.

D. W. HEYS & D. R. STUMP

Application of the Green's-function Monte Carlo
method to the compact Abelian lattice gauge theory

Phys. Rev. D **28**, 2067–2075 (1983)

The first application of diffusion or Green's function QMC in treating lattice gauge theories was reported in this paper on the compact U(1) gauge theory, one of the simplest of the lattice gauge theories. Earlier investigations of the U(1) and other Abelian gauge theories had been made by a variety of other methods, including perturbation expansions, variational calculations, and most successfully path integral treatments using Monte Carlo integrations. Aside from investigation of QMC methods for the U(1) theory, one object of this study was to verify results for the path integral approach in the Hamiltonian formulation. The Hamiltonian itself was expressed as a summation of terms involving electric and magnetic fields at each site, link, or plaquette, for which the authors developed the usual Green's function expression in a discrete form to be applied to an $n(p)$-space wavefunction $X[n(p)]$, a function of integer-valued plaquette variables. Except for discrete spatial positions, the Green's function calculations were very nearly the same as those for electronic structure studies.

The systems treated were small lattices, 5×5 and $3 \times 3 \times 3$, in two and three dimensions with a variable coupling constant λ. Variational calculations with different trial functions suitable for strong and weak coupling were found to reproduce known behavior for strong and weak limits. The Green's function calculations were facilitated with importance sampling and produced similar results, thus indicating the high quality of the trial wavefunctions. The results provide evidence for the existence of a phase transition in the three-dimensional case and for the nonexistence of a phase transition in the two-dimensional case.

28

V. R. PANDHARIPANDE,
J. G. ZABOLITZKY, S. C. PIEPER,
R. B. WIRINGA, & U. HELMBRECHT

Calculations of ground-state properties of liquid ^{4}He droplets

Phys. Rev. Lett. **50**, 1676–1679 (1983)

Earlier calculations[a] of the binding energy and compressibility of liquid helium in bulk were successful in reproducing experimental measurements within a few percent. These indicated a pairwise-additive empirical potential to be entirely adequate. In this paper the same potential was used in similar variational and Green's function QMC calculations to predict the properties of the ground states of liquid ^{4}He droplets, or more recently "clusters," containing 4 to 728 atoms. The calculation procedure was identical to that of the earlier work[a] except that boundaries were eliminated. Trial wavefunctions for both types of calculations were expressed as products of functions of interatomic distances similar to those used for the infinite liquid. A comparison of energies for the two types of calculations over the range of 4 to 112 atoms, showing the Green's function energies only 2–3% lower than the variational energies, indicates a high accuracy for the variational functions by themselves. Density distributions obtained as long-term averages by both methods were somewhat noisy at the droplet centers, but smooth and in good agreement away from the centers. These were fitted by quadratic and higher order functions from which surface tension could be estimated. For the larger droplets, the density was constant across the central region and approached the density of the bulk liquid. The authors noted that the success of these calculations gives confidence in the use of liquid drop formulas in predicting properties of nuclear matter.

[a]M. H. Kalos, M. A. Lee, P. A. Whitlock, and G. V. Chester, Green's function Monte Carlo method for liquid ^{3}He, *Phys. Rev. Lett.* **46**, 728 (1981).

M. A. LEE, P. VASHISHTA, & R. K. KALIA

Ground state of excitonic molecules by the Green's-function Monte Carlo method

Phys. Rev. Lett. **51**, 2422–2425 (1983)

The biexciton bound complex of two electrons and two holes is equivalent to the hydrogen molecule with different masses[a] and it can be treated in the same way as the hydrogen molecule by QMC methods. For equal electron and proton masses the complex is equivalent to a positronium molecule. The authors applied simple diffusion QMC with importance sampling to calculate ground-state energies of biexcitons for a range of electron-to-hole mass ratios $\sigma = m_e/m_h$. The full four-body problem was treated with positive and negative charges of unity interacting with Coulomb potentials. Since the ground state is nodeless, the problem of nodes does not occur, and the importance sampling function was taken as a simple product of electron-electron and hole-hole Jastrow functions with exponential terms. The results gave energies considerably lower than variational estimates and better agreement with experimental measurements for several crystalline materials. The limiting behavior as $\sigma \to 0$ was found in excellent agreement with the exact result.

[a] An early study: E. A. Hylleraas and A. Ore, Binding energy of the positronium molecule, *Phys. Rev.* **71**, 493 (1947).

30

F. MENTCH & J. B. ANDERSON

Quantum chemistry by random walk: Linear H_3

J. Chem. Phys. **80**, 2675–2680 (1984)

This is first of three papers published in 1984 which report calculations for the system H-H-H in a symmetric linear configuration matching the expected saddle point for one of the most basic chemical reactions, the exchange reaction $H + H_2 \rightarrow H_2 + H$. In this case, the calculations were fixed-node diffusion calculations, with importance sampling made with four different types of trial functions specifying the nodal structure. Each included simple electron-electron correlation terms. The functions were optimized by minimizing the variance in the local energy $H\psi_T/\psi_T$ using a simplex routine. The lowest energy result obtained for the minimum in the barrier to reaction was -1.6582 ± 0.0003 hartrees, corresponding to a barrier height of 10.2 ± 0.2 kcal/mol relative to the exact minimum for separated $H + H_2$. This value was slightly below the upper bound of 10.28 kcal/mol established in analytic variational calculations and about 0.6 kcal/mol above the expected value. The difference was attributed to node location error.

31

D. M. CEPERLEY & B. J. ALDER

Quantum Monte Carlo for molecules: Green's function and nodal release

J. Chem. Phys. **81**, 5833–5844 (1984)

This paper describes the application of the nodal release or "transient estimate" method,[a] successfully used earlier for treating the electron gas problem, to several small molecules: H_3, LiH, Li_2, and H_2O. This method can, in principle, provide a means for overcoming the node location problem for such systems, but it does require difficult extrapolations for most systems of interest. The basic calculations are carried out as for simple Green's function QMC without nodes as boundaries. A trial wavefunction ψ_T is used to locate approximate node positions, and a guide function ψ_G, everywhere positive and approximately equal to $|\psi_T|$ and nonzero but small at the nodes, is used for importance sampling. A released-node calculation is begun with walkers at steady state in a fixed-node calculation. These are then released and permitted to cross fixed-node surfaces. After a number of steps, the distribution of walkers includes walkers inside and outside the starting region. The difference (inside less outside) approaches the fermion ground state distribution, but with increasing number of steps the difference is lost in the noise as the inside and outside approach boson ground state distributions. Nevertheless, useful results were obtained for the systems examined, with energies correct within their statistical uncertainties. For the largest system, H_2O, the calculated energy was -76.43 ± 0.02 hartrees, which may be compared with a value of -76.438 hartrees based on experimental measurements.

[a]D. M. Ceperley and B. J. Alder, Ground state of the electron gas by a stochastic method, *Phys. Rev. Lett.* **45**, 566 (1980).

32

P. J. REYNOLDS, R. N. BARNETT, & W. A. LESTER, JR.

Quantum Monte Carlo study of the classical barrier height for the H + H_2 exchange reaction: Restricted versus unrestricted trial functions

Int. J. Quantum Chem., Symp. Ser. **18**, 709–717 (1984)

In this fixed-node calculation with importance sampling for the H-H-H system, the authors investigated the use of several single-determinant SCF trial functions and the effects of restricting the orbital coefficients. The true wavefunction must have identical coefficients for α and β electrons, but in variational calculations, the use of unrestricted spin-orbitals (allowing different coefficients) leads to energies lower than (or at least as low as) those for restricted spin-orbitals (requiring identical coefficients). This has been observed in variational QMC calculations for H-H-H. However, in fixed-node diffusion calculations, in which the energies depend only on node locations, the authors found a lower energy for restricted trial functions. For the barrier they obtained an energy of -1.6590 ± 0.0004 hartrees, a value lower than the lowest CI energy at the time, and only 9.69 ± 0.25 kcal/mol above that for separated reactants. This value is an upper bound to the barrier height, but within its uncertainty it overlaps the best recent value[a] of 9.608 ± 0.001 kcal/mol.

[a] K. E. Riley and J. B. Anderson, Higher accuracy quantum Monte Carlo calculations of the barrier for the H + H_2 reaction, *J. Chem. Phys.* **118**, 3437 (2003).

33

R. K. KALIA, P. VASHISHTA, & M. A. LEE

Binding energy of positively charged acceptors in
germanium — A Green's function Monte Carlo
calculation

Solid State Comm. **52**, 873–876 (1984)

Spectroscopic measurements of germanium crystals have revealed some of the characteristics of impurity complexes which can
be explained in terms of charge acceptors interacting with surrounding holes to form so-called pseudoatoms. With suitable modifications, these complexes can be treated in the same way as real
atoms and ions by conventional quantum methods, as described in
earlier work, and by QMC methods as described first in this paper. The complexes of interest include double acceptor impurities
Be and Zn binding three holes and quadruple acceptors binding
four holes, equivalent to negatively charged infinite-mass nuclei (of
charge $-Z$) surrounded by positively charged holes (of charge $+1$).
With use of effective masses for the holes, an effective dielectric
constant, and a modification of Fermi statistics to allow four holes
in $1s$ orbitals, these systems may be treated as simple atoms and
ions in diffusion QMC with importance sampling.

The systems treated were equivalent to nodeless H^-, He^+, He,
He^-, Li^+, Li, and Li^-. Both variational and diffusion calculations used the same trial wavefunctions, which were symmetrized
$1s$ orbital products with added correlation terms linear in the interelectron distances. For the one- and two-hole systems, the energies
could be deduced from the analogous systems. For these the diffusion QMC calculations gave equivalent results. In all cases the
diffusion calculations gave slightly lower energies than the variational calculations. In comparisons with experiment, agreement
was good for double acceptors and poor for triple acceptors, for
which the pseudoatom model was judged to be inadequate.

34

P. J. REYNOLDS, M. DUPUIS, & W. A. LESTER, JR.

Quantum Monte Carlo calculation of the singlet-triplet splitting in methylene

J. Chem. Phys. **82**, 1983–1990 (1985)

The methylene molecule CH_2 is one of those simple species with a long history in the development of quantum chemistry. First observed experimentally in 1959 by Herzberg and Shoosmith[a] it was the object of an especially important near-SCF study by Foster and Boys[b] predicting (correctly) the ground state to be the triplet $^3B_1$ with a bond angle of about 129°. Shortly thereafter, Herzberg concluded the molecule in the $^3B_1$ state was linear, but eventually experiments by others, as well as higher level calculations, showed Foster and Boys to be correct.[c] The singlet-triplet splitting (singlet energy minus triplet energy), initially thought to be about 2 kcal/mole from experiment and about 30 kcal/mol from theory, is now known to be 9–11 kcal/mole.

The fixed-node calculations of Reynolds et al. were carried out for optimum geometries indicated by analytic variational calculations, and they gave energies more than 15 kcal/mole below the lowest energy variational calculations available at the time. The trial functions for node location and importance sampling were single-determinant double-zeta functions with optimized Jastrow terms. The QMC result for singlet-triplet splitting, obtained by subtracting energies, is 9.4 ± 2.2 kcal/mole. With allowance for the zero-point energy and relativistic effects the splitting becomes 8.9 ± 2.2 kcal/mole. These are in very good agreement with very recent theoretical and experimental values. This substantial cancellation of node-location error for these two similar systems occurs in the same way as expected for the somewhat larger errors in analytic variational calculations for similar systems.

[a] G. Herzberg and J. Shoosmith, Spectrum and structure of the free methylene radical, *Nature* **183**, 1801 (1959).

[b] J. M. Foster and S. F. Boys, Quantum variational calculations for a range of CH_2 configurations, *Rev. Mod. Phys.* **32**, 305 (1960).

[c] For the full story see: H. F. Schaefer III, Methylene — A paradigm for computational quantum chemistry, *Science* **231**, 1100 (1986).

35

J. B. ANDERSON

Quantum chemistry by random walk: A faster algorithm

J. Chem. Phys. **82**, 2262–2663 (1985)

This little paper describes two procedures for achieving higher speeds in fixed-node diffusion QMC calculations with importance sampling. The first is an iterative procedure for reducing time-step error, thereby allowing larger and fewer time steps to obtain a given accuracy. The movement of a walker is divided into two parts: a diffusion step selected at random from an appropriate distribution and a drift step calculated from the drift velocity at the initial location. These give a temporary step to a new location for which the drift velocity is again determined. The average of the two velocities together with the original diffusion step is used to determine a permanent step. In tests, the procedure was found to reduce time-step error by a factor of 5 to 10. The second procedure makes use of two trial functions for importance sampling, the first a simple function for walker motion and the second a more complex and more accurate function for determining energies. These allow the more computationally expensive function to be used at large intervals for samples having a lower serial correlation.

36

D. F. COKER, R. E. MILLER, & R. O. WATTS

The infrared predissociation spectra of water clusters

J. Chem. Phys. **82**, 3554–3562 (1985)

Coker, Miller, and Watts carried out experimental measurements of the infrared predissociation spectra of small water clusters in molecular beams formed with free jet expansions of water vapor in helium. The spectra include several O-H stretch absorptions and an H-O-H bend overtone. Theoretical analyses based on empirical intramolecular/intermolecular potential energy surfaces were made using normal mode theory, local mode theory, and a novel extension of the simple random-walk procedure without importance sampling. The normal mode analysis was found inadequate and the local mode analysis not quite satisfactory, but the QMC-based quantum simulation procedure predicted the observed vibrational bands for the dimer and trimer very accurately. The simple QMC method was used to generate a large number of walkers with a distribution corresponding to the ground state wavefunction ψ_0. This many-body function was then projected onto the local vibrational coordinates for each molecule to give single-variable functions, each then fit by a Morse function. A variational calculation including cross terms for low-lying excited states was then made with this reference state. The results gave unambiguous assignments for the observed O-H stretching vibrations in the dimer, good agreement for the bending overtones, and satisfactory agreement for the O-H vibrations in the trimer. Repetition using a different pair interaction potential established that the experiment could provide a sensitive method for testing pair interactions.

37

D. CEPERLEY & B. J. ALDER

Muon–alpha-particle sticking probability in muon-catalyzed fusion

Phys. Rev. A **31**, 1999–2004 (1985)

A negative muon μ has the negative charge of an electron but a mass 207 times that of an electron. Like an electron e binding two protons p in the molecular ion H_2^+ or $(epp)^+$, it can bind two protons in a molecular ion $(\mu pp)^+$, and because of its higher mass, it binds more strongly. Similarly, a negative muon can bind a deuteron d and a triton t in a molecular ion $(\mu dt)^+$. With the deuteron and triton held tightly together, they may fuse together, ejecting a neutron, forming an alpha particle, and releasing substantial energy. In about 1% of the events, the muon sticks to the positively charged alpha particle and is lost as a catalyst to additional fusions.

This paper describes the use of Green's function QMC to calculate the wavefunction for the $(\mu dt)^+$ molecular ion for the purpose of estimating the sticking probability of the muon with the alpha particle. The essential functions are the initial and final wavefunctions at the instant of fusion with the deuteron and triton at the same position (coalescence point), and these functions along with the sudden approximation can be used to estimate the sticking probability. Earlier analytic variational treatments based on the Born-Oppenheimer approximation were not satisfactory. A modified version of Green's function QMC with the use of a trial density matrix, along with importance sampling, was used to determine energies and sample the wavefunction. Calculations of relative wavefunction values, at the coalescence point relative to others, were made using the evolution procedure described by Liu, Kalos, and Chester.[a] Determination of normalization factors required special consideration. The calculations were clearly successful and gave a sticking probability of 0.90%, a value about 25% smaller than given by a Born-Oppenheimer wavefunction and closer to estimates derived from experimental measurements.

[a]K. S. Liu, M. H. Kalos, and G. V. Chester, Quantum hard spheres in a channel, *Phys. Rev. A* **10**, 303 (1974).

38

R. J. HARRISON & N. C. HANDY

Quantum Monte Carlo calculations on Be and LiH

Chem. Phys. Lett. **113**, 257–263 (1985)

In their first paper in the quantum Monte Carlo area, Harrison and Handy investigated the use of multi-configuration trial wavefunctions and various forms of Jastrow functions in fixed-node QMC. Calculations were repeated for the small molecules He, Be^{2+}, Be, and LiH. For the nodeless ground states of He and Be^{2+}, they used two- to six-term Hylleraas functions and found significant gains in efficiency for the more accurate trial functions despite the additional terms. They reproduced known energies for these species within statistical error. For Be, it was found that the fixed-node results for a five-term multi-configuration SCF function were far superior to those for a single-term SCF trial function. For LiH, the use of one-, two-, and four-term MC-SCF functions was found to produce little improvement in the variance in local energies or the values of the energies obtained. Recovery of correlation energy was about 99% in each case. This paper introduced the use of higher order terms in a Jastrow function similar to those used by Handy and Boys[a] in their transcorrelated method. Also considered was the use of an additional electron-nucleus term in the Jastrow function. These attempts at improving the Jastrow factor were not very successful, but planted the seeds for later successes by Schmidt and Moskowitz.[b]

[a] N. C. Handy and S. F. Boys, A calculation for the energies and wavefunctions for states of neon with full electron correlation accuracy, *Proc. Roy. Soc. (London) A* **310**, 63 (1969); **311**, 309 (1969).

[b] K. E. Schmidt, J. W. Moskowitz, Correlated Monte-Carlo wave-functions for the atoms He through Ne, *J. Chem. Phys.* **93**, 4172 (1990).

B. H. WELLS

The differential Green's function Monte Carlo method.
The dipole moment of LiH

Chem. Phys. Lett. **115**, 89–94 (1985)

Use of correlated random walks in fixed-node diffusion QMC offers the opportunity to calculate differences in energies of similar systems. In this paper Wells reports such calculations for LiH in the presence of differing field strengths to determine its dipole moment. Since the local energies for identical electron configuration and importance sampling functions differ only in the potential energy due to the applied electrical field, the local energies are strongly correlated, and the statistical error in the difference in energies can be much lower than that for either of the individual energies. The calculations were started from a ψ_T^2 distribution of walkers with identical moves for each system. Since the multiplication terms were slightly different, variable weights were used for each walker. Although these eventually diverge for the correlated pairs, there is a sufficient period of strong correlation to allow efficient reduction in variance in energy differences. Thus, for an applied field of 0.0001 a.u. versus no field, the energies obtained were -8.0593 ± 0.0030 hartrees and -8.0595 ± 0.0030 hartrees, but the computed energy difference was 0.000227 ± 0.000003 hartrees, corresponding to a dipole moment of 2.27 ± 0.03 au, in good agreement with the value derived from experiments. Since there was no provision for a change in the node structure with applied field, it appears the dipole moment is insensitive to the change if any occurs.

40

J. W. MOSKOWITZ & K. E. SCHMIDT

The domain Green's function method

J. Chem. Phys. **85**, 2868–2874 (1986)

The boundary conditions required in fixed-node QMC are not appropriate for the use of Green's functions in a simple way. In this paper, Moskowitz and Schmidt describe modifications to allow the direct use of Green's functions without the short-time approximation, i.e., without the discretization of the diffusion equation. In this modified procedure, walkers are allowed to move only within a domain in which the trial function specifying the nodes is positive. A linear approximation is used to predict the distance to a node, and the step size is revised accordingly. The distance to a region of net positive potential is similarly considered. The method was used in fixed-node calculations for LiH and H_2O with SCF trial functions, as well as for Be and Be-H_2 with MCSCF trial functions. Recovery of correlation energy was better than 90% for the best trial functions. The method was developed to supplant the fixed-node diffusion method, with an accompanying "proof" that the fixed-node diffusion methods of earlier workers (Anderson,[a] Reynolds et al.,[a] and presumably Fermi) were incorrect. The proof contained a mistake subsequently reported[b] by the authors.

[a] J. B. Anderson, *J. Chem. Phys.* **63**, 1499 (1975); **65**, 4121 (1976); **73**, 3897 (1982); P. J. Reynolds, D. M. Ceperley, B. J. Alder, and W. A. Lester, Jr., *J. Chem. Phys.* **77**, 5593 (1982).

[b] J. W. Moskowitz and K. E. Schmidt, *J. Chem. Phys.* **87**, 1906 (1987).

41

D. F. COKER & R. O. WATTS

Quantum simulation of systems with nodal surfaces

Mol. Phys. **58**, 1113–1123 (1986)

Coker and Watts reported modifications of the simple diffusion QMC algorithm to treat the problem of node location in small systems. This enabled the calculation of low-lying states for a vibrating diatomic molecule as well as electronic structures for H $2p$, He ^3S, and Li ^1S. The general method makes use of cancellation of positive and negative walkers occupying the same volume elements of a grid in the coordinate space of the walkers, together with appropriate constraints to prevent decay to a lower-energy solution. In the case of molecular vibration, orthogonality to lower states was enforced by maintaining equal weights of positive and negative walkers in equivalent regions of coordinate space for lower states. For the case of fermion ground states for He ^3S and Li ^1S, equal weights of positive and negative walkers were required for all coordinate space. For the higher dimensionality of the Li case, cancellation was based on proximity of walkers of opposite signs. Results in terms of energy were accurate to about three digits and within their uncertainties overlapped accepted values for each of the systems investigated. These provided a hint of more accurate results to be obtained later for small systems.

42

P. J. REYNOLDS, R. N. BARNETT, B. L. HAMMOND, R. M. GRIMES, & W. A. LESTER, JR.

Quantum chemistry by quantum Monte Carlo: Beyond ground-state energy calculations

Int. J. Quantum Chem. **29**, 589–596 (1986)

The paper describes several extensions of fixed-node diffusion in the areas of obtaining analytic derivatives of the energies of molecules, calculating expectation values for quantities other than the energy, and treating excited states.

The calculation of energy derivatives is illustrated with a calculation of the first derivative of the energy of the ground state of H_2 with respect to the internuclear distance, without using finite differences. With importance sampling the energy is obtained from a weighted average of local energies expressed in analytic form. These may be differentiated to obtain their derivatives with respect to any of several variables, including the internuclear distance, and their weighted averages may be used to determine the desired derivatives. The results for the H_2 energy derivatives were in agreement with independent determinations.

For calculation of expectation values of the energy, sampling of the (true wavefunction)(trial wavefunction) $\psi\psi_T$ distribution is sufficient, but other quantities usually require sampling the ψ^2 distribution, which is not so readily available in QMC calculations. Extrapolations of ψ_T^2 and $\psi\psi_T$ results, from variational and diffusion calculations, to ψ^2 results are illustrated for various expectation values for the case of ground-state H_2.

The third item illustrated is the use of the fixed-node calculations for fermion systems in excited states having the same symmetry as their ground states. The examples are He ($1s2s\ {}^1S$) and H_2 ($B\ {}^1\Sigma_u^+$), with nodes specified by two-determinant wavefunctions. The calculated energies were somewhat higher than the known values for the species, but the trial functions were not optimized, and the results were sufficiently accurate to illustrate the effectiveness of the approach.

43

G. SUGIYAMA & S. E. KOONIN

Auxiliary field Monte Carlo for quantum many-body ground states

Ann. Phys. **168**, 1–26 (1986)

This paper describes a different QMC method for solving the Schrödinger equation, one unlike variational, diffusion, and Green's function methods. It can be applied to many-body systems to produce solutions for ground states. For systems of fermions, antisymmetrization can be imposed to overcome the problem of node location, but the instability of the method is a serious shortcoming, so that the method as described is limited to treating relatively simple systems. Nevertheless, this auxiliary field quantum Monte Carlo (AFMC) method was demonstrated to be successful for test cases reported in this paper, and in time has been improved to allow treatment of realistic systems.[a]

The method makes use of the imaginary-time propagation of an initial wavefunction toward the ground-state function. It requires use of a trial wavefunction which is the product of single-particle orbitals for bosons or an antisymmetrized product for fermions. A Hubbard-Stratonovich transformation of the propagator changes the many-body problem to a problem of a collection of single particles interacting with a fluctuating (auxiliary) field which replaces the many-body interactions. The authors developed an integral expression for the energy which, after discretization, could be evaluated by a Metropolis Monte Carlo integration forward in time. The algorithm was tested for one-dimensional systems of particles: 6 to 20 bosons with varied interactions, 2 to 6 neutron-proton pairs, others with repulsive potentials. Importance sampling was found to be critical and computational requirements to be large. For spin systems on a lattice the AFMC method was considered feasible, but for general multidimensional systems the method appeared to be limited by its computation requirements.

---

[a]R. Baer, Ab initio computation of molecular singlet-triplet energy differences using auxiliary field Monte Carlo, *Chem. Phys. Lett.* **343**, 535 (2001).

44

M. M. HURLEY & P. A. CHRISTIANSEN

Relativistic effective potentials in quantum Monte Carlo
calculations

J. Chem. Phys. **86**, 1069–1070 (1987)

This is the first of three papers,[a] each from a different group,
to report the use of effective potentials, model potentials, or pseudopotentials in QMC calculations in 1987 and 1988. The authors
point out the advantages of eliminating core electrons in order
to simplify the calculations, reduce the number of singularities in
the wavefunction, and thereby reduce the variance in local energies. They note the primary difficulty in using relativistic effective
potentials is the nonlocal character of most potentials and their
dependence on angular projection operators rather than on mere
electron configurations. In this study, the authors defined a simple
local potential V in terms of the nonlocal effective potentials oper-
ating on a trial wavefunction. The extent of the error introduced
was unknown but possibly small. The effective potentials were
taken directly from earlier analytic studies for the species considered. The trial wavefunctions were single determinants of Gaussian
orbitals together with Jastrow correlation terms. Calculations were
carried out for Li/Li$^-$ and K/K$^-$ so that electron affinities could be
obtained. The values so determined were 0.56 and 0.52 eV, respectively, compared to 0.618 and 0.502 eV determined in experiments.
It was noted that corresponding all-electron calculations for the
F/F$^-$ pair required a vastly greater computation effort, perhaps
more than a factor of 1000 greater.

[a]The others are: B. L. Hammond, P. J. Reynolds, and W. A. Lester, Jr.,
Valence quantum Monte Carlo with ab initio effective core potentials, *J. Chem.
Phys.* **87**, 1130 (1987) and T. Yoshida and K. Iguchi, Quantum Monte Carlo
with the model potential, *J. Chem. Phys.* **88**, 1032 (1988).

45

D. R. GARMER & J. B. ANDERSON

Quantum chemistry by random walk: Methane

J. Chem. Phys. **86**, 4025–4029 (1987)

The methane molecule has served as a test case for a lengthy series of electronic structure calculations beginning in 1941 with approximate Hartree-Fock calculations and an energy of -39.47 hartrees, reaching in 1973 to 81% of the correlation energy with extended-basis CI calculations at -40.4584 h, and going much further with this paper in 1987 to 97% of the correlation energy with fixed-node diffusion QMC calculations at -40.506 h. The calculations were carried out by the fixed-node diffusion method with drift, making use of importance sampling with a single determinant trial function based on double-zeta Slater-type orbitals, along with a simple Jastrow function. This function was optimized to minimize both the energy and the variance of the local energy. Time-step error was investigated in detail with six different time-step sizes down to 0.0005 h^{-1}, giving nearly straight line behavior. The calculated energy at $-40.506(2)$ hartrees and 97% of the correlation energy was 31 kcal/mol lower than the lowest analytic variational energy at the time and only 5 kcal/mol above the experimental nonrelativistic energy.

46

D. F. COKER & R. O. WATTS

The diffusion Monte Carlo method for quantum systems
at nonzero temperatures

J. Phys. Chem. **91**, 4866–4873 (1987)

The Schrödinger equation in imaginary time is not the only
equation which is isomorphic to the diffusion equation. One important example is the Bloch equation of quantum statistical mechanics,

$$\frac{\partial F}{\partial \beta} = \sum_i \frac{\hbar^2}{2m_i} \nabla_i^2 F - V(\mathbf{r})F.$$

This equation may be solved for F, the Fourier transform of the
equilibrium density matrix and, with additional analysis, it gives
the probability of finding a quantum system at thermal equilibrium at temperature $1/\beta$ in a volume element at a point $\mathbf{r}$ in a
multidimensional configuration space. With an initial distribution
of walkers representing F selected from a classical distribution at
high temperature, the diffusion/birth/death procedure for simple
diffusion QMC may be followed in time-equivalent β to determine
distributions at lower temperatures. Many of the devices used for
single-state calculations, including importance sampling, may be
adopted without modification.

Coker and Watts developed these basic algorithms and demonstrated the accuracy of the method for the one-dimensional harmonic oscillator and for the O-H dimer treated as a Morse oscillator. They went on to generate quantum distributions for a
low-density neon gas (two atoms in a periodic box) and the water dimer (confined to prevent dissociation). In each of these four
cases, comparisons with results from alternate methods (such as
matrix squaring, evaluation of Slater sums, and use of Feynman
paths) showed excellent agreement and indicated the "general applicability of the random walk algorithm."

47

B. L. HAMMOND, P. J. REYNOLDS, & W. A. LESTER, JR.

Valence quantum Monte Carlo with *ab initio* effective core potentials

J. Chem. Phys. **87**, 1130–1136 (1987)

This paper reports a more elaborate treatment than either of the other papers reporting effective potentials in QMC calculations in 1987 and 1988. Although justified here as a means to avoid the higher computational requirements of heavy (large Z) atoms, the advantages of eliminating core electrons apply equally well for systems of lighter atoms such as carbon, nitrogen, and oxygen. The treatment in this paper begins with a valence-electron Schrödinger equation in which core electrons are missing. The nuclear charges are replaced by reduced effective nuclear charges, and the core-valence electron-electron repulsions are replaced by a nonlocal effective core potential U^{ECP}. This term is a relatively complicated expression which incorporates the core-valence orthogonality condition and is a function of valence electron positions. It is obtained in the course of Hartree-Fock calculations for the atoms involved. The implementation in QMC calculations required a number of minor approximations, along with a lengthy set of derivations, finally resulting in an expression for U^{ECP} compatible with QMC.

Calculations were carried out in fixed-node diffusion QMC for several systems: Li/Li^-, Na/Na^+, Mg/Mg^+, $NaH/(Na+H)$, and $Na_2/(Na+Na)$. These produced electron affinities, ionization potentials, and binding energies which could be compared with experimental values. Statistical uncertainties were typically 0.02 eV for these values, and all agreed with experimental values within their uncertainties. Such accuracies would have required much greater computational efforts for all-electron calculations.

48, 49, 50

S. M. ROTHSTEIN & J. VRBIK

A Green's function used in diffusion Monte Carlo

J. Chem. Phys. **87**, 1902–1903 (1987)

J. B. ANDERSON & D. R. GARMER

Validity of random walk methods in the limit of small time steps

J. Chem. Phys. **87**, 1903–1904 (1987)

P. J. REYNOLDS, R. K. OWEN, & W. A. LESTER, JR.

Is there a zeroth order time-step error in diffusion quantum Monte Carlo?

J. Chem. Phys. **87**, 1905–1906 (1987)

These three notes each offer evidence against an earlier claim, added as an appendix to a paper by Moskowitz and Schmidt,[a] that the discretization of the imaginary-time Schrödinger equation was in error even for infinitesimal time-step sizes and that all results with this short-time approximation were incorrect. As acknowledged by Moskowitz and Schmidt, there was an error in their analysis and they agreed that the short-time approximation was correct after all.

[a] J. W. Moskowitz, K. E. Schmidt, The domain Green's function method, *J. Chem. Phys.* **85**, 2868 (1986); **87**, 1906 (1987).

51

D. F. COKER & R. O. WATTS

Diffusion Monte Carlo simulation of condensed systems

J. Chem. Phys. **86**, 5703–5707 (1987)

This paper reports diffusion quantum Monte Carlo calculations for condensed systems, in this case for liquid ^{4}He and for solid molecular H_2. Calculations for liquid He made earlier by Whitlock et al.[a] using Green's function QMC for the same potential energy surface allowed a comparison of results for this system. Coker and Watts used spherical pair potentials derived from scattering experiments for both systems, along with Axilrod-Teller-Muto three-body terms. The calculations were performed with importance sampling, using a simple product of pair functions as a trial function for helium. For hydrogen a simple pair function multiplied by a single-particle term for position in the fcc lattice was used. Periodic boundary conditions were employed, and calculations were carried out for several particle densities of each species. For helium, good agreement with the prior Green's function calculations was obtained for thermodynamic and structural properties and for radial distribution functions. For hydrogen, the calculated average energies and pressures gave general agreement with experimental measurement, but they differed up to 10% in energy and 12% in pressure in a manner to be expected for neglect of molecular anisotropy.

[a] P. A. Whitlock, D. M. Ceperley, G. V. Chester, and M. H. Kalos, Properties of liquid and solid ^{4}He, *Phys. Rev. B* **19**, 5598 (1979); **21**, 999 (1980).

52

S. M. ROTHSTEIN, N. PATIL, & J. VRBIK

Time step error in diffusion Monte Carlo simulations:
An empirical study

J. Comput. Chem. **8**, 412–419 (1987)

The discretization of the differential equations describing the transient behavior of a system usually introduces a time-step error. The problem occurs in diffusion QMC and may, in principle, be overcome by varying the time-step size and extrapolating results to a step size of zero. This paper shows that such extrapolations are not so simple, and it provides explanations for some of the surprises encountered in using several different procedures. Following the importance sampling procedure of Grimm and Storer,[a] the Green's function and similarly discretization may be separated into individual functions or terms for drift, diffusion, and branching (or weight change). Rothstein et al. combined these in varied orders of full and half steps and examined their behavior in calculations for H_2 and LiH with different time step sizes. Both linear $E = E_0 + a\Delta t$ and quadratic $E = E_0 + a(\Delta t)^2$ behaviors were observed, and even these depended on the system. Further analysis gave insights into the causes for the trends observed and some practical advice for selecting the best procedures.

[a]R. C. Grimm and R. G. Storer, Monte-Carlo solution of Schrödinger's equation, *J. Comput. Phys.* **7**, 134 (1971).

53

G. J. MARTYNA & B. J. BERNE

Structure and energetics of Xe_n^-

J. Chem. Phys. **88**, 4516–4525 (1988)

The problem treated in this paper is that of understanding xenon cluster ions at the microscopic level. Some properties of cluster anions had been measured, but these revealed no detailed information on structure and energetics. In the study reported here, diffusion QMC was used to investigate electron binding to small clusters of up to 19 atoms in fixed minimum-energy configurations. This was followed by path integral calculations to determine equilibrium distributions of cluster structures at temperatures of 10–100 K for several cluster sizes.

Diffusion QMC calculations for frozen clusters were carried out for an additive electron-xenon pseudopotential: first, using a trial wavefunction given by the product of s-wave solutions for single electron-xenon pairs with the pseudopotential interaction, and second, with an improved function based on the results of the first and minimization of the energies. Variational energies could be determined by numerical quadrature for this three-dimensional problem.

For clusters of five and fewer atoms, the binding energy of the electron could not be determined, although an upper bound was found to be 12 K. For 6–19 atoms the binding energy increased in monotonic fashion from 12 K to 1100 K. The change in binding energy with number of atoms exhibited variations which could be explained in terms of the cluster structures. For the smaller clusters, the electron probability distribution was found diffuse and extended outside the cluster. The path integral calculations gave good agreement with the diffusion QMC calculations for the smaller clusters. For larger clusters of 50 and 100 atoms, the electron distribution was localized within the cluster.

54

T. YOSHIDA & K. IGUCHI

Quantum Monte Carlo method with the model potential

J. Chem. Phys. **88**, 1032–1034 (1988)

This third paper on the subject of effective potentials in 1987 and 1988 reports a somewhat simpler approach to combining effective core potentials with QMC than other papers of the same era. In this work the effective potential is based on an improved version of one introduced by Bonifacic and Huzinaga,[a] but it is simplified by elimination of core-valence exchange interaction and by use of an approximate core-valence orthogonality. All that remains of the original Hamiltonian are the kinetic energy terms for valence electrons, the potential energy terms for valence electrons interacting with each other and a reduced-charge nucleus, and a simpler radius-dependent potential energy term for each valence electron. Fixed nodes are introduced with a Slater-determinant trial function also used for importance sampling.

Tests were made in calculations of the valence energies of the atoms Mg, Ca, and Sr, along with their cations, so that ionization potentials for these atoms could be calculated and compared with experimental values. The statistical uncertainties in the energies were about 0.03 eV, and the deviations in ionization potentials from experiment were about 0.1 eV. The results were significantly better than those of SCF calculations exhibited for comparison. Additional calculations were reported in a later paper.[b]

---

[a]V. Bonifacic and S. Huzinaga, Atomic and molecular calculations with the model potential method. I, *J. Chem. Phys.* **60**, 2779 (1974).

[b]T. Yoshida, Y. Mizushima, and K. Iguchi, Electron affinity of Cl: A model potential-quantum Monte Carlo study, *J. Chem. Phys.* **89**, 5815 (1988).

55

M. CAFFAREL & P. CLAVERIE

Development of a pure diffusion quantum Monte Carlo
method using a full generalized Feynman-Kac formula.
I and II

J. Chem. Phys. **88**, 1088–1099, 1100–1109 (1988)

These two papers describe the development of a variant of diffusion QMC with importance sampling and its applications to a few
simple systems. The method is based on the generalized Feynman-
Kac path integral formalism[a] and is termed the "full generalized
Feynman-Kac" or FGFK method. Its major advantage, relative
to other diffusion-based methods, is the elimination of branching
or multiplication of walkers. Its major disadvantage is a greater
complexity.

The basic implementation is very nearly the same as that of
standard diffusion QMC with importance sampling, and it incorporates diffusion with drift in the same way. The uses of trial
functions, fixed nodes, extrapolation to zero time-step size, and
other procedures except for branching are identical. However, the
evaluation of mean quantities can be carried out in terms of ψ^2
sampling to obtain energies and other properties based on the
FGFK formula. The applications described include ground and
excited states of the harmonic oscillator and the helium atom and
the ground state of the hydrogen molecule. For these the energies
as well as a dozen other observables such as the averages of interelectron distances were determined. The results were found to be
in good agreement with accepted values. A combination of FGFK
with the released-node method was investigated for the case of the
one-dimensional harmonic oscillator and found to be successful for
that case.

[a] M. Caffarel and P. Claverie, Treatment of the Schrödinger-equation
through a Monte-Carlo method based on the generalized Feynman-Kac formula, *J. Stat. Phys.* **43**, 797 (1986).

56

D. R. GARMER & J. B. ANDERSON

Potential energies for the reaction $F + H_2 \rightarrow HF + H$
by the random walk method

J. Chem. Phys. **89**, 3050–3056 (1988)

If the reaction $H + H_2 \rightarrow H_2 + H$ is number one on the list of interesting chemical reactions, then number two is the reaction $F + H_2 \rightarrow HF + H$. It is very fast, it is highly exothermic, it drives chemical lasers, it can be studied in crossed molecular beams, it can be studied by infrared spectroscopy, and it involves only 11 electrons. This paper, and its companion paper[a] a year earlier, report fixed-node diffusion QMC calculations of the ground-state potential energy surface for the F-H-H systems, with emphasis on the region of the barrier to reaction. At the time this work was completed, it was clear from experiments that the barrier height must be in the range of 1 to 2 kcal/mol. Calculations by a variety of methods gave barrier heights in the range of 1 to 6 kcal/mole for collinear reactions. More recent high-level calculations, QMC and others, place the collinear barrier at about 3 kcal/mol, with a lower height of about 2 kcal/mol for slightly bent configurations.

The QMC calculations were fixed-node diffusion calculations with importance sampling, using single-determinant trial functions with orbitals represented by cubic spline functions fit to Gaussian orbitals. Recovery of correlation energy was about 96% for the isolated F atom and for the HF molecule. Total energies for reactions and products were about 40 kcal/mol lower than the lowest energy variational calculations at the time, and only about 10 kcal/mol above energies derived from experiment. The calculated exothermicity was 29.0 ± 1.4 kcal/mol, and that derived from experiment was 31.7 kcal/mol. The barrier height was found to be 4.5 ± 0.6 kcal/mol for collinear reaction. A calculation for a single bent configuration indicated a slight decrease in height for bending. The calculated barrier height was thought by the authors to be 1–3 kcal/mol higher than compatible with experiment, and it was suggested that node location error was responsible.

[a]D. R. Garmer and J. B. Anderson, Quantum chemistry by random walk: Application to the potential energy surface for $F + H_2 \rightarrow HF + H$, *J. Chem. Phys.* **86**, 7237 (1987).

57

S. FAHY, X. W. WANG, & S. G. LOUIE

Variational quantum Monte Carlo nonlocal pseudopotential approach to solids: Cohesive and structural properties of diamond

Phys. Rev. Lett. **61**, 1631–1634 (1988)

This paper describes one of the earliest applications of QMC techniques to solids. The calculations are variational calculations for the ground state of diamond, with comparison calculations for the carbon atom. The study introduces nonlocal pseudopotentials originally developed for density functional calculations at the LDA level and makes use of LDA single-particle wavefunctions in constructing the trial function. The trial wavefunction ψ_T is of the standard single-determinant Slater-Jastrow type for the valence electrons of the carbon atoms, adapted for a solid structure. The integrals required for evaluating the nonlocal potentials and the associated interactions contributing to the energy were calculated for each configuration using a simple but effective sampling procedure. Metropolis-style variational calculations with sampling based on ψ_T^2 were carried out for 16 atoms (64 electrons) and 54 atoms (216 electrons) in diamond-lattice supercells with periodic boundary conditions. Energies were evaluated using Ewald sums. The results, for binding energy and for the lattice constant, gave excellent agreement with experiment. The energy minimum was found for a lattice constant within about 1% of the measured value, and the cohesive energy calculated for that lattice was 7.45(7) eV per atom, which is about one standard deviation from the experimental value of 7.37 eV per atom. This represents a significant improvement over the corresponding LDA calculation, which predicts 8.63 eV per atom.

58

C. J. UMRIGAR, K. G. WILSON, & J. W. WILKINS

Optimized trial wave functions for quantum Monte Carlo calculations

Phys. Rev. Lett. **60**, 1719–1722 (1988)

Although many of the diffusion QMC and variational QMC calculations of electronic structure used minimization of the variance in local energies to optimize trial functions in the decades before this paper was published, it provides useful examples of the advantages of the procedure. Even the first electronic structure calculations with importance sampling for molecular systems[a] used trial wavefunctions optimized in this way for a fixed set of configurations. The paper reports optimizations of various combinations of determinants with Jastrow and exponential Padé functions for five two-electron systems from H^- to Be^{2+} and for the four-electron Be atom. For the two-electron systems, the standard deviations in local energies could be reduced to a few millihartrees, and the errors in total energies could be reduced to as low as one microhartree. This corresponds to a recovery of more than 99.99% of the correlation energy. In the case of the Be atom, a four-determinant function with an exponential Padé correlation term and a total of 47 free parameters gave a standard deviation in local energy of 0.09 hartrees, a total energy only 1 millihartree above the "exact" energy, and recovery of 99% of the correlation energy. In later work, minimizations of combinations of variance and energy values for optimization have yielded slightly better results in some cases.

[a]F. Mentch and J. B. Anderson, Quantum chemistry by random walk: Importance sampling for H_3^+, *J. Chem. Phys.* **74**, 6307 (1981).

59

B. L. HAMMOND, P. J. REYNOLDS, & W. A. LESTER, JR.

Damped-core quantum Monte Carlo method: Effective treatment for large-Z systems

Phys. Rev. Lett. **61**, 2312–2315 (1988)

Heavy atoms are problematic in QMC calculations for several reasons. Their energies are large, they have many electrons, and the time scales for core and valence electron motions are widely different. Larger local energy fluctuations are associated with higher energies. More electrons require longer computational efforts. Different time scales require short steps in one region and long equilibration times in others. The result is an estimated scaling of computational effort with Z^{6-7}, the sixth to seventh power of the nuclear charge Z. The Z-dependence can be reduced with the use of pseudopotentials, and this approach has been pursued with success, but not without introducing uncertainties. This paper describes an entirely different method which retains all the electrons, but treats core and valence electrons in different ways.

The basic idea is that of treating core electrons by Metropolis sampling and valence electrons by diffusion with drift. Separate trial functions $\psi_{T,\text{core}}$ and $\psi_{T,\text{val}}$ were chosen orthogonal such that the valence electrons were prevented from collapsing into the core. For the core electrons the Metropolis walks were taken without branching and produced a $\psi^2_{T,\text{core}}$ distribution of walkers. For the valence electrons normal diffusion walks with drift produced a $\psi_{\text{val}}\psi_{T,\text{val}}$ distribution. A smooth transition between regions was obtained with a damping function to eliminate branching as valence electrons approached the core.

The method was applied to the atoms C (2 core and 4 valence electrons), Si (10 core and 4 valence electrons), and Ge (28 core and 4 valence electrons). The effectiveness of the method was demonstrated with calculations of ionization potentials and electron affinities within 0.1 eV of experimental values for each of these atoms. The reduction in computational requirements was as high as a factor of 5000 for Ge.

60

J. VRBIK, M. F. DePASQUALE, & S. M. ROTHSTEIN

Estimating the relativistic energy by diffusion quantum Monte Carlo

J. Chem. Phys. **88**, 3784–3787 (1988)

This is the first paper to report the use of QMC in estimating relativistic corrections to nonrelativistic energies of atoms and molecules. The corrections described are based on first-order perturbation theory without modification of the nonrelativistic wavefunctions, but higher order corrections might be similarly treated. The method is that of fixed-node diffusion QMC with importance sampling and without multiplication (branching), so that the distribution of walkers produced corresponds to ψ_T^2, the square of the trial wavefunction. The authors note that incorporating multiplication would improve the accuracy but require more difficult calculations.

The corrections determined include those known as (a) the (nonrelativistic) Hughes-Eckart correction, (b) the term for the relativistic mass effect, (c) the Darwin term, and (d) Breit's term. Spin-orbit interactions, spin-spin interactions, and many lesser effects are usually not important and are not included. Each of the four corrections is calculated as an average of quantities which are functions of ψ_T, $\nabla\psi_T$, $\nabla^2\psi_T$, the potential energy V, ∇V, and $\nabla^2 V$. The corrections for the relativistic mass term and the Darwin term tend to cancel each other, are strongly correlated, and may be calculated simultaneously to reduce statistical error.

The method is illustrated with calculations for the molecule LiH using a trial wavefunction giving a nonrelativistic variational energy of -8.020 ± 0.006 hartree. (The exact nonrelativistic energy was recently estimated to be -8.070553 h.) The computed estimates of the corrections are (a) $+0.00302$ h, (b) $-0.00351(3)$ h, (c) $+0.00279(5)$ h, (b + c) $-0.00072(2)$ h, and (d) -0.00003 h, with indicated statistical error. Systematic error, largely due to uncertainty in the trial wavefunction, is estimated to be up to 10% of the values. The total relativistic correction (b + c + d) is calculated to be $-0.00073(2)$ h. Its value is similar to that given by a relativistic SCF calculation close to the Hartree-Fock limit.

61

J. CARLSON

Alpha particle structure

Phys. Rev. C **38**, 1879–1885 (1988)

The time-independent Schrödinger equation for the case of nucleons in light nuclei is more difficult to treat than that of electrons in a light atom because of more complicated interaction terms. Nevertheless, for three-nucleon systems such as the triton, binding energies and structural details have been successfully obtained for realistic nucleon-nucleon interaction models by several different methods, especially by analytic variational methods. For four-nucleon systems such as the alpha particle, realistic models of the interactions are also available, but the added nucleon makes the treatments more difficult and the properties predicted less certain. This paper reports the use of variational QMC calculations to obtain accurate solutions for the alpha particle with several nucleon-nucleon interaction models. For one model, both variational and diffusion QMC calculations provided the opportunity for direct comparisons to assess the accuracy of variational QMC results.

The interaction models, chosen to provide a test of the sensitivity of results, included those known as Argonne V14, Nijmegen, Bonn, Reid V8, and Urbana Model 7. Trial wavefunctions previously developed for such problems were constructed of Slater determinants, functions of internucleon distances and spins, as well as correlation terms. Optimizations of parameters were carried out by minimizing the expectation values of the energies, making use of correlated sampling. Metropolis sampling was used for the variational QMC calculations. For the diffusion QMC calculations, the time-step was varied for extrapolation to zero step size. Energies were obtained directly, but expectation values depending on the square of the wavefunction were extrapolated from ψ_T^2 to $\psi\psi_T$ to ψ^2 values. The results allowed a comparison of a number of properties for the different models with accuracies similar to those available for three-nucleon systems. The utility of the diffusion QMC calculations was clearly demonstrated by revealing several significant differences from the variational results.

62

C. A. TRAYNOR & J. B. ANDERSON

Parallel Monte Carlo calculations to determine energy
differences among similar molecular structures

Chem. Phys. Lett. **147**, 389–394 (1988)

This study combines correlated sampling with Green's function sampling in QMC for highly accurate calculations of potential energy differences among similar molecular structures. The method is exact for nodeless systems. It is illustrated with calculations for the molecular ion H_3^+ to determine the minimum-energy structure and the potential energy surface in that region. The key to Green's function sampling is the use of geometrically similar structures all related to a primary structure by a single length parameter. The movement of walkers in the secondary system (or systems) is geometrically similar to that in the primary, the electron configurations are geometrically similar, and the sampling of positions is completely correlated. Since the potential energies are not the same for secondary walkers, the multiplication terms V/E differ, and the weights of secondary walkers diverge from those of the primary walkers as the calculations proceed. Division of secondary walkers with high weights, and elimination of those with low weights, is matched to that of the primary systems. Importance sampling may be incorporated with a single trial function or with scaled trial functions.

The results for H_3^+ showed the minimum energy structure for the equilateral triangle to be very close to 1.6500 bohr in side length. The calculations for 12 different structures at the same time required about the same calculation effort as for a single structure when scaled trial functions were used. Energy differences were obtained with high accuracy for structures differing by up to about 10% in size. For those differing more than about 10%, the accuracies in energy differences were much lower.

63

M. CAFFAREL, P. CLAVERIE, C. MIJOULE, J. ANDZELM, & D. R. SALAHUB

Quantum Monte Carlo method for some model and realistic coupled anharmonic oscillators

J. Chem. Phys. **90**, 990–1002 (1989)

This paper reports a first attack on the node location problem for excited states of two-dimensional systems corresponding to coupled one-dimensional oscillators. The zero-point energy of such systems may be readily evaluated using diffusion QMC with or without importance sampling, but as for electronic states, the problem of node locations prevents such easy solutions for excited vibrational states unless the nodes can be specified from symmetry considerations.

The authors present a modified fixed-node approach which is a good approximation when the nodes of a reference function are close to the exact nodes. The method is based on variations in node locations to minimize the energies in the regions separated by nodes, subject to the constraint that the energies must be equal. This is accomplished with a complete wavefunction made up of a variable combination of wavefunctions corresponding to those for uncoupled oscillators. Applications for several two-dimensional systems including a realistic model for adsorbed CO on a surface gave essentially exact agreement with the results of analytic variational solutions with large basis sets. The authors concluded that their method avoided many of the problems encountered with other methods and appeared well adapted for systems of strongly coupled oscillators and multiple-well potentials.

64

J. CARLSON, J. W. MOSKOWITZ, & K. E. SCHMIDT

Model Hamiltonians for atomic and molecular systems

J. Chem. Phys. **90**, 1003–1006 (1989)

The aim of this study was to reduce the variance in the energies for QMC calculations for atomic and molecular systems by replacing the electron-electron potential energy terms $1/r_{ij}$ for core electrons by averaged values. The complete Hamiltonian expression was thus altered to become a model Hamiltonian, and while total energies from such calculations may have little meaning, energy differences for similar systems obtained with the same model may give fairly good agreement with those from calculations for the full Hamiltonian. The work reported in this paper indicated that is the case for Li, LiH, and Li_2. The average electron-electron potential was expressed as a two-parameter function decreasing with increasing distance from a nucleus. Importance sampling was carried out with a trial function consisting of an unaltered Slater determinant specifying node locations multiplied by a modified Jastrow function. The random walk of configurations was executed with Green's function sampling for the full set of electrons. Energy differences for LiH versus (Li + H) and Li_2 versus (Li + Li) were accurate within about 1 millihartree. The reduction in variance in local energies led to increases in calculation efficiency by a factor of 5 to 7 for these systems. The results indicated such an approach might be successful for larger systems.

65

V. DOBROSAVLJEVIĆ, C. W. HENEBRY, & R. M. STRATT

Simulation of the electronic structure of an atom dissolved in a hard-sphere liquid

J. Chem. Phys. **91**, 2470–2478 (1989)

For this paper the authors used diffusion Monte Carlo methods to obtain a first solution to the problem of electronic structure of an atom in a structured liquid. Although models of several types using the assumption of no spatial variation of solvent effects have had some success, they miss essential features which are included naturally in a QMC calculation. The authors treated the problem of a hydrogenic atom in a hard-sphere liquid, specifically a proton and an electron in the presence of a hard-sphere helium-atom solvent. A separate classical simulation of 31 helium atoms and 1 hydrogen atom was used to generate equilibrium configurations of the solvent for 300 K. These were followed by diffusion QMC calculations for the electron in the field of the fixed proton, along with the boundary conditions of a zero wavefunction at the surfaces of helium spheres. A highly effective importance sampling function, incorporating factors of the Jastrow form, reduced the uncertainties in calculated energies and simplified the analysis to determine the characteristics of the electronic structure. Results were obtained for thousands of configurations for each of a number of conditions for liquid structures as well as for bcc structures. The major observations were a smooth increase in ground-state energy with liquid density and a solvent-induced hybridization of the p orbital into the ground state at intermediate densities.

66

V. MOHAN & J. B. ANDERSON

Effect of crystallite shape on exciton energy: Quantum
Monte Carlo calculations

Chem. Phys. Lett. **156**, 520–524 (1989)

About the time that small semiconductor crystallites were first
called "quantum dots", quantum Monte Carlo calculations were
first applied in studies of these systems. Small crystallites have
spectroscopic properties like those of molecules with molecular orbitals, and they shift to the behavior of electrons and holes in continuous bands as their size is increased. This paper reports a study
of the effect of size and shape of crystallites on ground-state energies for an electron-hole pair (or Wannier exciton) confined to the
crystallite. The system was investigated using a model by Brus[a]
with the electron-hole pair similar to a hydrogen atom having effective masses for each, unit charges for each, and a coulombic interaction governed by a bulk dielectric constant. The electron and hole
were confined to the crystallite by setting the charge densities to
zero at the boundaries. Estimates of the energies for several sizes of
spherical CdS and ZnO crystallites had been made earlier in variational calculations by Brus[a] and by others. The calculations of this
later paper were carried out for spheres, cubes, and rectangular parallelopipeds of CdS and ZnO crystallites using diffusion QMC with
importance sampling. The trial function was a product of three-
dimensional particle-in-a-box (or sphere) functions for electron and
hole, multiplied by a hydrogen-like function of the electron-hole distance. For small crystallites the effect of confining the electron and
hole in close proximity was a significant lowering of energy below
that for a noninteracting system. Results for the spherical crystallites matched closely those of variational calculations. Energies
were obtained for parallelopipeds of widely varying shapes — corresponding to quantum dots, quantum wires, and quantum wells
— as well as for one- and two-dimensional hydrogen-like systems.
The behavior of all of these could be explained in terms of the relative importance of confinement by boundaries and the interaction
of the electron and hole.

[a]L. E. Brus, A simple model for the ionization potential, electron affinity,
and aqueous redox potentials of small semiconductor crystallites, *J. Chem.
Phys.* **79**, 5566 (1983); **80**, 4403 (1984).

67

G. SUGIYAMA, G. ZERAH, & B. J. ALDER

Ground-state properties of metallic lithium

Physica A **156**, 144–168 (1989)

As reported in this paper, both variational and fixed-node diffusion QMC calculations were used to determine electronic properties for bulk lithium. The calculations were carried out for cubic supercells containing lithium nuclei at fixed sites in body-centered cubic and face-centered cubic lattice structures with up to 108 atoms. The trial functions for importance sampling were constructed in two forms: one with Gaussian plane-wave functions and the other with more complex augmented plane-wave functions. Slightly lower energies were obtained in variational QMC for the Gaussian form, and equivalent results were obtained in fixed-node diffusion QMC for the two forms. Simple spherical trial functions gave significantly higher energies for compressed systems. Energies, pair distribution functions, and electron densities were investigated for crystal densities ranging from about 0.25 to 90 times the densities at normal pressures. The variation of cohesive energy with density was found in reasonable agreement with experimental results in calculations using the larger numbers of atoms in the supercell. The lattice constants were also in good agreement for these cases. In general, this paper showed that fixed-node diffusion calculations were feasible for metallic systems and could be competitive with alternative methods.

68

H. SUN & R. O. WATTS

Diffusion Monte Carlo simulations of hydrogen fluoride dimers

J. Chem. Phys. **92**, 603–616 (1990)

This is one of the earlier QMC papers to describe calculations of molecular vibrational states. As noted by the authors, harmonic oscillator models for vibration are not adequate for treating floppy molecules, and the principal motivation for their study was to "give up the harmonic oscillator assumption and use quantum Monte Carlo to calculate accurate vibrational frequencies for the HF and DF dimers." Clearly, they were successful in meeting that objective. In doing so they devised several new approaches to the general problem of treating vibrational motion.

Diffusion QMC with importance sampling was used with an improved short-time approximation based on an added multiplication term second order in the time-step size. Excited states for which node locations could be specified from consideration of symmetry arguments, as for torsion states, were simulated in fixed-node calculations. For other states an orthogonalization method[a] based on cancellation of positive and negative walkers as well as orthogonality with lower states was used. For these it was necessary to simulate the excited states successively, and calculations were limited to four different states. The states involved in tunneling splitting for the dimers were distinguished by an added fixed node.

The calculations were carried out for three different potential energy surfaces and results were compared with spectroscopic data. Each one gave reasonable values for low-lying vibrations. Modification of one surface gave still better values. The calculations revealed strong mixing between intermolecular vibration and the bond-bending modes which was found sensitive to the choice of surface. The tunneling splitting could be calculated accurately and was also found sensitive to the choice of surface. For systems such as these, with tunneling splittings and anharmonic vibrations, the QMC method was expected to be "particularly valuable."

[a]D. F. Coker and R. O. Watts, Structure and vibrational spectroscopy of the water dimer using quantum simulation, *J. Phys. Chem.* **91**, 2513 (1987).

69

V. MOHAN & J. B. ANDERSON

Quantum Monte Carlo calculations of three-body
corrections in the interaction of three helium atoms

J. Chem. Phys. **92**, 6971–6973 (1990)

For internulcear distances of about 4 bohr and greater, the potential energy of three helium atoms is very close to pairwise additive. That is, the total interaction energy is very close to the sum
of the interaction energies of the three pairs, each in the absence of
a third atom. The energy may then be conveniently expressed as

$$E = \sum_3 E_1 + \sum_3 \Delta E_2 + \Delta E_3,$$

where E_1 is an atom energy, ΔE_2 a pairwise interaction energy, and
ΔE_3 is a three-body correction term. For pairwise additivity, ΔE_3
is zero. The correction ΔE_3 is extremely small and is certainly
less than 30 microhartrees (about 10 K) for three He atoms in
an equilateral triangle configuration of side length 5.6 bohr with
a total energy of about 8.7 hartrees. A third-order perturbation
analysis by Axilrod and Teller[a] and Muto[b] (ATM) has provided
a simple and convenient expression for the three-body correction.
Clearly, it must fail at small internuclear distances, but it often has
been assumed to be acceptable for distances near the equilibrium
dimer distance of 5.6 bohr.

The fixed-node diffusion calculations reported were carried out
for equilateral triangles with side lengths of 2.5 to 6.5 bohr. The
trial function was based on Hylleraas expressions for the atoms
in a 36-term antisymmetric sum. The results confirmed that the
three-body correction term was less than 10 K for distances greater
than about 5.6 bohr, but the uncertainties were about 10 K. For
distances less than 5.6 bohr, the correction was larger and the ATM
expression found to be widely different. At 3.5 bohr the QMC value
was $-247(100)$ K and the ATM value 7.3 K. Remarkably, there was
good agreement of QMC, SCF, MP3, and CISD (but not ATM)
with each other within their uncertainties, at all distances.

---

[a]B. M. Axilrod and E. Teller, Interaction of the van der Waals type between three atoms, *J. Chem. Phys.* **11**, 299 (1943).

[b]Y. Muto, *Proc. Phys.-Math. Soc. Japan* **17**, 629 (1943).

70

K. E. SCHMIDT & J. W. MOSKOWITZ

Correlated Monte Carlo wave functions for the atoms He through Ne

J. Chem. Phys. **93**, 4172–4178 (1990)

This paper introduced an expanded form of correlation function to replace the simple Jastrow functions used in trial wavefunctions for both VQMC and DQMC calculations. The result was a significant decrease in the variance in local energies and an increase in recovery of correlation energies for the atoms examined (and for many other atoms and molecules, by subsequent workers using the new functions). The form for the trial functions employed is a single determinant multiplied by a correlation function F proposed by Boys and Handy[a]

$$F = \exp\left(\sum_{I,i<j} U_{Iij}\right),$$

$$U_{Iij} = \sum_{k}^{N(I)} \Delta(m_{kI}, n_{kI}) c_{kI} (\bar{r}_{iI}^{m_{kI}} \bar{r}_{jI}^{n_{kI}} + \bar{r}_{jI}^{m_{kI}} \bar{r}_{iI}^{n_{kI}}) \bar{r}_{ij}^{o_{kI}},$$

where the sums are over nuclei I, electrons i and j, and terms k up to $N(I)$. The $\bar{r}$ functions are of the type $\bar{r} = br/(1 + br)$ for both electron-electron and electron-nucleus interactions. The electron-nucleus terms have the important effect of compensating for electron density changes due to the electron-electron terms, and they give a large improvement over the authors' previous results. Functions with 7, 9, and 17 terms were optimized for each of the atoms He through Ne and used in variational QMC calculations. Recovery of correlation energy was approximately 100% for He, 97% for Li, and 18% to 85% for Be through Ne. In the case of Ne, the single-determinant with the correlation function recovered 85%, a value higher than the 75% obtained in configuration interaction calculations (with 107 determinants) at the time.

[a]S. F. Boys and N. C. Handy, A condition to remove indeterminacy in interelectronic correlation functions, *Proc. Roy. Soc. (London), Ser. A* **309**, 209 (1969).

71

M. V. RAMA KRISHNA & K. B. WHALEY

Wave functions of helium clusters

J. Chem. Phys. **93**, 6738–6751 (1990)

This paper describes improved trial wavefunctions for helium clusters and their use in variational QMC calculations of their binding energies, structural properties, and breathing mode vibrational excited states, for clusters of 3 to 270 atoms. The wavefunctions developed were similar to those of earlier work but included convenient analytic functions for two- and three-body correlations. All terms were of the Jastrow exponential form, based on sums of one-, two-, and three-body terms $t_1(r_i)$, $t_2(r_{ij})$, and $t_3(r_{ij}, r_{jk}, r_{ki})$, expressed as functions of interparticle distances. Excited-state wavefunctions for the breathing vibrational mode were obtained with the aid of procedures adapted from treatment of the quantum mechanics of liquid drop models. The interactions of helium atoms were specified as pairwise-additive with an accurate semi-empirical pair potential. Optimizations of parameters in the trial wavefunctions were carried out for individual cluster sizes with an exhaustive search procedure in a series of trial variational calculations, in which the parameters were varied to minimize the energy. A partial correlation of results was obtained by use of identical initial coordinates and random number seeds in these optimization calculations. Once the wavefunctions were optimized, longer calculations were used to determine properties of the clusters. In addition, diffusion QMC calculations were used to obtain more accurate nonvariational energies for comparisons.

For three- and four-atom clusters the structures were found to be delocalized, with triangular and tetrahedral structures favored. For larger clusters the structures were described as liquid-like. The diffusion QMC energies for the smaller clusters gave good agreement with earlier results. The fairly high accuracies of the wavefunction expressions were indicated by diffusion QMC energies found only slightly lower than the variational QMC energies.

72

S. FAHY, X. W. WANG, & S. G. LOUIE

Variational quantum Monte Carlo nonlocal
pseudopotential approach to solids: Formulation and
application to diamond, graphite, and silicon

Phys. Rev. B **42**, 3503–3522 (1990)

This paper describes the use of nonlocal pseudopotentials with variational QMC to calculate the binding energies of the solids diamond, graphite, and silicon. While nonlocal pseudopotentials have the advantage of elimination of core electrons and the large fluctuations of local energies associated with core electrons, they produce several types of difficulties in diffusion and Green's function QMC. As shown in this paper, the use of variational QMC avoids many of these. The calculations used pseudopotentials of the standard form developed for use in density functional calculations. The trial functions were based on mixtures of short-range Gaussian orbitals and long-range plane-wave functions in single determinants multiplied by Jastrow functions. Metropolis sampling was carried out for movement of one electron at a time, along with its images, in order to take advantage of the efficiencies associated with changing one row of the determinant for a single electron. The solids were treated as cells of up to 54 atoms (216 valence electrons) with periodic boundary conditions. For each of the three systems — diamond, graphite, and silicon — the VQMC structural properties were found in excellent agreement with experiment. Density functional calculations at the LDA level were similarly successful. More important, the VQMC binding energies and energy differences were found in excellent agreement with experiment. For these quantities density functional calculations at the LDA level were far less successful.

73

X. W. WANG, J. ZHU, S. G. LOUIE, & S. FAHY

Magnetic structure and equation of state of bcc solid hydrogen: A variational quantum Monte Carlo study

Phys. Rev. Lett. **65**, 2414–2417 (1990)

Solid hydrogen in its atomic bcc structure provides an interesting testing ground for various theoretical methods in predicting the properties of its several electronic states. These include three possible magnetic phases: paramagnetic (PM), ferromagnetic (FM), and antiferromagnetic (AFM). The system is simple enough to permit reliable calculations for all three phases to determine their energies as functions of the density and to predict the transitions between phases. The calculations reported in this paper were variational QMC calculations performed primarily on a 54-atom cluster in the bcc structure with periodic boundary conditions. The protons were fixed in position. The trial wavefunction was constructed from a product of spin-up and spin-down determinants based on single-particle wavefunctions from density-functional calculations for the system, multiplied by a symmetric Jastrow factor. The specification of magnetic phase was made through modifications of the single-particle orbitals within the determinants and localization of orbitals to one or the other of the cubic sublattices. The calculations for all three phases were performed for a range of densities in which earlier calculations had predicted magnetic and metal-insulator transitions. At the lowest densities, the FM and AFM energies approached the value for free hydrogen atoms, with AFM lower than FM. With increasing density the PM phase drops below the AFM and becomes the ground state. Across the range, the ferromagnetic phase was never the lowest-energy phase. Corrections applied to correct for finite-size effects of the cluster gave no change in the relative stabilities of the phases.

74

C. A. TRAYNOR, J. B. ANDERSON, & B. M. BOGHOSIAN

A quantum Monte Carlo calculation of the ground state energy of the hydrogen molecule

J. Chem. Phys. **94**, 3657–3664 (1991)

Since the wavefunction for the ground state of the hydrogen molecule has no nodes, its energy can be determined in QMC without node location error. This paper reports such calculations for the full four-body (two-electron, two-proton) problem to obtain the energy without the use of the Born-Oppenheimer approximation or any other adiabatic approximations. The calculations were carried out with diffusion QMC and with Green's function QMC. Importance sampling was incorporated into each of the calculations. For the diffusion calculations, extrapolation of energies obtained at several different time-step sizes was required. For the Green's function calculations there was no time-step error, and sign changes for walkers were avoided by use of a potential energy offset and a minor approximation to eliminate positive potential energies. The energies so determined were not exact but had extremely small systematic errors.

The large differences in electron and proton masses led to extensive computation requirements due to the slow equilibration and serial correlation induced by the heavier protons. The calculations were executed on one of the first massively parallel computers, a Thinking Machines CM-2 with 65,536 processors. Frequent communication among the processors was required to balance the number of walkers treated in each.

The Green's function results were more accurate than the diffusion results and had an energy of $-1.164\ 024 \pm 0.000\ 009$ hartrees, corresponding to a dissociation energy of $36117.9 \pm 2.0\ \mathrm{cm}^{-1}$, corrected for relativistic and radiative effects. The best experimental value available for comparison was about a factor of ten more accurate, at $36118.1 \pm 0.2\ \mathrm{cm}^{-1}$. The best values available from alternative theoretical predictions were also a factor of ten more accurate at 36117.9 to $36118.1\ \mathrm{cm}^{-1}$. Improvements in all of these surfaced a few years later.

75

M. QUACK & M. A. SUHM

Potential energy surfaces, quasiadiabatic channels, rovibrational spectra, and intramolecular dynamics of $(HF)_2$ and its isotopomers from quantum Monte Carlo calculations

J. Chem. Phys. **95**, 28–59 (1991)

The hydrogen fluoride dimer $(HF)_2$ is the simplest molecule exhibiting the phenomenon of a hydrogen bond. In this case, the hydrogen fluoride molecules are loosely held together by the relatively weak hydrogen bond to produce a floppy molecule with anharmonic vibrations not easily characterized, but producing an interesting and complex spectroscopy. In this paper, Quack and Suhm report the use of the diffusion QMC method in analyzing the behavior of the dimer and its isotopomers, and the successful prediction and assignment of spectral bands observed in their earlier experiments.[a]

The DQMC calculations were carried out for nuclear motion on two different six-dimensional potential energy surfaces adjusted to reproduce experimentally measured spectra including anharmonic interactions between all vibrational modes. Two new QMC methods for calculating excited rotational and vibrational states were introduced: a clamped-coordinate quasiadiabatic channel and a centrifugal-energy approximation scheme. In general, quantitative agreement with experimental observations was obtained, and the success of DQMC methods in the prediction and analysis of the spectra of hydrogen-bonded systems was clearly demonstrated.

[a]M. Quack and M. A. Suhm, Observation and assignment of the hydrogen-bond exchange disrotatory in-plane bending vibration ν_5 in $(HF)_2$, *Chem. Phys. Lett.* **171**, 517 (1990).

76

L. MITÁŠ, E. L. SHIRLEY, & D. M. CEPERLEY

Nonlocal pseudopotentials and diffusion Monte Carlo

J. Chem. Phys. **95**, 3467–3475 (1991)

This paper reports improvements which overcome some of the problems encountered in the use of nonlocal pseudopotentials in QMC. These increase the reliability of the method by reducing or eliminating several approximations of earlier treatments, reduce the variance in local energies, and increase the recovery of correlation energy. Calculations for Si, Sc, and Cu neutrals and ions in several states and the dimer Si_2 were used to test the revised method.

One key improvement was the use of correlated trial wavefunctions making use of Schmidt-Moskowitz expressions for two- and three-body terms. Additional exponential terms were added in some cases. Atomic orbitals for the determinants were obtained in density functional calculations. Another key improvement was in the procedure for integrating the trial function over a sphere in determining the electron-pseudopotential energetics. For the complete correlated trial function the integrals were determined with numerical quadrature, settling on 18-point quadrature with random orientation on the sphere. Except for these changes, the calculations were similar to those of the earlier studies. Both variational and fixed-node diffusion calculations were made.

The recovery of valence correlation energy for atoms and ions was 80%-90% in the variational calculations and estimated to be greater than 95% in the diffusion calculations. Comparison of ionization potentials and electron affinities with experimental values showed agreement within 1% for Si, a few percent for Sc, and about 20% for Cu. The general conclusion was that the method was stable and feasible even for the difficult case of Cu.

77

S. A. ALEXANDER, R. L. COLDWELL, H. J. MONKHORST, & J. D. MORGAN, III

Monte Carlo eigenvalue and variance estimates from several functional optimizations

J. Chem. Phys. **95**, 6622–6633 (1991)

Optimization of trial functions by minimization of the variance in local energies began with early variational calculations by Frost,[a] Conroy,[b] and later by Coldwell.[c] Minimizing the variance in local energies clearly leads to VQMC and DQMC energies with a lowered statistical error, but as pointed out in this paper, there may be better choices of functionals to optimize. Ten different functionals, assembled from local energy and weighting terms, all giving emphasis in varying degrees to the variance in local energies and the average of local energies, were investigated, with low-lying states of He, H_2, and H_3^+ as example systems. The trial wavefunctions were simple compact functions including explicit r_{12} terms, with as many as 36 parameters total for He ^3S and 122 for H_3^+. Optimization for a fixed set of configurations was followed by VQMC calculations for the most promising trial functions generated. Minimizing the variance alone in the optimizations gave the lowest variance in energies for the VQMC results. Minimizing a combination of energy and variance in a form used by Conroy[b] gave the lowest energies in most but not all cases. The choice of functional was found to have a significant effect on the accuracy of VQMC calculations, and the best choice was found to be system dependent.

[a] A. A. Frost, The approximate solution of Schrödinger equations by a least squares method, *J. Chem. Phys.* **10**, 240 (1942).

[b] H. Conroy, Molecular Schrödinger equation. I. One-electron solutions, *J. Chem. Phys.* **41**, 1327 (1964).

[c] R. L. Coldwell and R. E. Lowther, Monte Carlo calculation of the Born-Oppenheimer potential between two helium atoms using Hylleraas-type electronic wave functions, *Int. J. Quantum Chem., Symp. Ser.* **12**, 329 (1978).

78

J. B. ANDERSON, C. A. TRAYNOR, & B. M. BOGHOSIAN

Quantum chemistry by random walk: Exact treatment
of many-electron systems

J. Chem. Phys. **95**, 7418–7425 (1991)

This paper reports the first exact quantum treatment of a many-electron system. The QMC method used is exact in that it requires no mathematical approximations and no physical approximations beyond those of the Schrödinger equation. It requires no interpolations and no extrapolations. As in most Monte Carlo methods, there is a statistical or sampling error which is readily estimated and which may be made arbitrarily small by additional computation. The method is illustrated with applications to the problems of a particle in a two-dimensional box, excited H_2 $^3\Sigma_u^+$, and linear symmetric H-H-H. The "sign problem in fermion Monte Carlo" is solved by a cancellation scheme in which positive and negative walkers in close proximity may cancel each other, a device mentioned in an early paper by Ulam[a] and described later by Arnow et al.[b] Time-step error is eliminated by use of Green's function QMC, and cancellation is made on the basis of overlapping Green's functions for movement of walkers, made efficient by a Monte Carlo procedure to eliminate conditional acceptance of walker moves. Importance sampling is incorporated in order to reduce statistical error in the energies determined.

For the case of clamped-nucleus H_2 $^3\Sigma_u^+$ at an internuclear distance of 1.4 bohr, the calculated energy was -0.7838 ± 0.0007 hartrees, a value consistent with prior analytic variational calculations. For the H-H-H system at separations of 1.757 bohr, the energy determined was -1.6591 ± 0.0003 hartrees, a value lower than that of any variational calculation, corresponding to a barrier height of 9.6 ± 0.2 kcal/mol.

[a]S. M. Ulam, *A Collection of Mathematical Problems,* Interscience Publishers, New York, 1960, pp. 123–128.

[b]D. M. Arnow, M. H. Kalos, M. A. Lee, and K. E. Schmidt, Green's function Monte Carlo for few fermion problems, *J. Chem. Phys.* **77**, 5562 (1982).

79

G. AN & J. M. J. VAN LEEUWEN

Fixed-node Monte Carlo study of the two-dimensional Hubbard model

Phys. Rev. B **44**, 9410–9417 (1991)

This study extended the fixed-node QMC method to a lattice model for hopping fermions using Green's function techniques. In general, the modifications of the procedures for continuous systems to adapt them to a lattice system were minor, and there appeared to be no surprises. Applications were made to the two-dimensional Hubbard model with different on-site interactions, lattice dimensions, boundary conditions, node specifications, and trial wavefunctions for importance sampling. Results were compared with those of Hartree-Fock and variational QMC calculations. For the HF calculations with zero interaction, the exact solutions are known, and the Green's function QMC results agreed with those within statistical error.

Trial wavefunctions specifying the nodes were obtained directly as those of HF determinantal solutions for zero interaction, or as variationally optimized determinantal functions with correlation terms. The procedure for fixed nodes was to eliminate walkers crossing the boundaries of a nodal region, a procedure not quite correct, but probably a reasonable approximation. The results in all cases were consistent with the HF and variational QMC calculations, giving lower energies than either for nonzero interaction energies. The existence of a domain-wall phase suggested by the HF and variational QMC results was confirmed by the fixed-node Green's function results. It was noted that the fixed-node method would be applicable to systems other than the Hubbard model.

80

R. N. BARNETT, P. J. REYNOLDS, & W. A. LESTER, JR.

Monte Carlo algorithms for expectation values of coordinate operators

J. Comput. Phys. **96**, 258–276 (1991)

In diffusion QMC with importance sampling, walkers provide samples of configurations with probabilities proportional to $\psi\psi_T$, the product of the true wavefunction ψ and a trial wavefunction ψ_T. The expectation value of the energy (very fortunately) may be calculated from the local energies of these samples as $E = \langle\psi H\psi_T\rangle/\langle\psi\psi_T\rangle = \langle\psi E_{loc}\psi_T\rangle/\langle\psi\psi_T\rangle$. But for an operator A which does not commute with the Hamiltonian H, the mixed expectation value $\langle\psi A\psi_T\rangle$ is only an approximation to the "pure" expectation value $\langle\psi A\psi\rangle$. Thus, for most quantities of interest other than the energy, an extension to standard diffusion QMC methods is required. This paper describes two approaches to the problem. Both are based on an earlier development by Liu, Kalos, and Chester[a] involving the tracking of descendents of walkers to obtain pure expectation values. The first uses simple DQMC and the second uses VQMC with DQMC "side walkers." These methods were tested in their application to the model systems H and H_2, and both were found effective in yielding accuracies and precisions correct within 0.5% for the quantities $\langle r\rangle$ for H and $\langle r^2\rangle$ and $\langle z^2\rangle$ for H and H_2.

[a]K. S. Liu, M. H. Kalos, and G. V. Chester, Quantum hard spheres in a channel, *Phys. Rev. A* **10**, 303 (1974).

81

M. CAFFAREL & O. HESS

Quantum Monte Carlo perturbation calculations of interaction energies

Phys. Rev. A **43**, 2139–2151 (1991)

The perturbation approach to predicting interaction energies between molecules avoids many of the problems associated with accurate calculations of total energies for subtraction to obtain energy differences, but it has many problems of its own. This paper reports a slightly successful first attempt to combine the perturbation approach with QMC methods for such calculations. The basis for the development is the nth-order Rayleigh-Schrödinger perturbation theory, which in principle can be formulated to give a series of energy corrections $\Delta E^{(n)}$ that can be expressed as integrals involving the perturbing potential and the unperturbed wavefunctions. In practice this is difficult for interactions of molecules, but the authors succeeded in reformulating the problem so that the required quantities could be determined in diffusion QMC calculations. The interaction energy was split into Rayleigh-Schrödinger and exchange parts which could be separately evaluated. The QMC aspects were carried out in diffusion QMC with importance sampling using Jastrow-like trial functions and conditional acceptance of moves to enforce detailed balance. This produced the expectation values required in the expressions for the interaction energy.

The combination method was tested in calculations of the He-He interaction at short internuclear distances. Typical values of the energy terms agreed with known values within a few percent, and statistical error was about 0.01 hartree. This is not sufficiently accurate to be useful in itself for predictions of the He-He interaction, but it is an indicator of a successful first step in the development of a new method.

82

X.-P. LI, D. M. CEPERLEY, & R. M. MARTIN

Cohesive energy of silicon by the Green's-function Monte Carlo method

Phys. Rev. B **44**, 10929–10932 (1991)

After variational QMC calculations with pseudopotentials for diamond-structure carbon and silicon, the next step higher is fixed-node diffusion QMC for these materials. This paper reports that step for silicon with use of a pseudo-Hamiltonian[a] to eliminate core electrons. In this case, the pseudo-Hamiltonian was tested and found accurate in fixed-node diffusion calculations for atomic and diatomic silicon. The trial wavefunction specifying the node locations for valence electrons was a Slater-Jastrow combination based on a plane-wave basis set with constants determined in LDA density functional calculations. Both VQMC and fixed-node diffusion QMC calculations were performed for supercells of 64 atoms (216 electrons) with a range of lattice constants. Corrections of results to infinite systems were made with the assumption of size dependence the same as for LDA calculations. The predicted minimum-energy structures for VQMC and FN-DQMC were approximately the same, with lattice constants within 1% of the experimental value. The cohesive energy for FN-DQMC was approximately 0.4 eV lower than for VQMC and, at 4.51(3) eV per atom, was within the uncertainties in agreement with experimental values of 4.63(8) eV per atom. Questions regarding the transferability of the LDA-based pseudo-Hamiltonian led the authors to recommend the use of pseudo-Hamiltonians or pseudopotentials based on many-body theory in future calculations. As noted by the authors, this work demonstrated that QMC calculations of solids can provide high accuracies.

[a]G. B. Bachelet, D. M. Ceperley, and M. G. B. Chiocchetti, Novel pseudo-Hamiltonian for quantum Monte Carlo simulations, *Phys. Rev. Lett.* **62**, 2088 (1989).

83

Z. SUN, R. N. BARNETT, & W. A. LESTER, JR.

Optimization of a multideterminant wave function for quantum Monte Carlo: Li_2 $(X\ ^1\Sigma_g^+)$

J. Chem. Phys. **96**, 2422–2423 (1992)

For many systems a single determinant Hartree-Fock wavefunction is not adequate for use as a trial function specifying the nodes for a fixed-node QMC calculation. For nondynamical correlation effects such as those occurring with a $2s-2p$ near degeneracy and for the dissociation of many molecules a multideterminant wavefunction is required. The minimal bases for such calculations may be discovered in low-level analytic variational calculations with configuration interaction. For this paper the authors investigated use of a four-determinant function for the $^1\Sigma_g^+$ ground state of the dimer Li_2 in both variational and fixed-node QMC.

The four determinants were assembled from four different σ molecular orbitals (from $1s, 2s$, and $2p_z$ AOs) and two π orbitals (from $2p_x$ and $2p_y$ AOs). The sum of the four determinants with four adjustable coefficients was multiplied by electron-electron and electron-nucleus Jastrow terms.

The results showed the advantages of multideterminant functions but also revealed some hazards in their use. For Li_2 at its equilibrium separation of 5.05 bohr, the optimization of 41 parameters, including the two free coefficients of the determinants, to minimize the variance in local energies gave the lowest VQMC energy (recovering 77% of the correlation energy). However, in FN-DQMC the lowest energy (recovering 100% of the correlation energy) was obtained with the coefficients of the determinants given by an original MCSCF calculation without optimization. Thus, minimization of the variance failed to optimize the node structure. Further, it was found that the complete optimization produced spurious nodes and in so doing raised the FN-DQMC energy obtained. Clearly, multideterminant wavefunctions must be used with care.

84

R. N. BARNETT, P. J. REYNOLDS, & W. A. LESTER, JR.

Computation of transition dipole moments by Monte Carlo

J. Chem. Phys. **96**, 2141–2154 (1992)

The interest in QMC for calculating transition dipole moments is driven by the fact that these terms are very sensitive to the quality of the wavefunctions from which they are calculated. Wavefunctions giving excellent energies may lead to very poor transition moments. In this paper the authors report an investigation of several different approaches within the QMC area that might prove useful in calculating transition moments.

The test case for comparisons is the transition dipole moment between the $1s$ and $2p_x$ states of the hydrogen atom for which the value is known. The transition moment $\mathbf{r}_{mn}$ is given by the expression

$$\mathbf{r}_{mn} = \langle \phi_m \mid \Sigma_i \mathbf{r}_i \mid \phi_n \rangle,$$

where the sum is over all electrons, so that sampling must be based on the product of the two wavefunctions $\phi_m \phi_n$. The authors devised three different ways to do this, with two based on VQMC and the third based on FN-DQMC. The last appears most successful. It requires a Green's function averaging approach with configurations for states m and n generated with probabilities proportional to $\psi_T \phi$, the product of trial and true wavefunctions, by independent guided walks. For points sampled from one distribution the Green's function is averaged for points from the other distribution. The combination leads to a distribution with weights proportional to $\psi_T \phi$ from which the transition dipole is calculated. As demonstrated this method appears quite reliable and it was deemed to hold promise for larger systems.

85

J. B. ANDERSON

Quantum chemistry by random walk: Higher accuracy for H_3^+

J. Chem. Phys. **96**, 3702–3706 (1992)

Although the ground-state Born-Oppenheimer wavefunction of the molecular ion H_3^+ is nodeless and cancellation of positive and negative walkers is not required to treat a node structure, the larger step sizes of "exact" cancellation calculations[a] in Green's function QMC made it the choice for the calculations reported in this paper. In this case the higher efficiencies allowed, for the first time ever, the determination of potential energies for a polyatomic system with microhartree accuracy.

The calculations were essentially identical to those described earlier[a] for H_2 $^1\Sigma_u^+$ and linear symmetric H-H-H. The calculated energy for H_3^+ in its equilibrium equilateral triangle configuration of side length 1.6500 bohr was found to be $-1.343\ 835 \pm 0.000\ 001$ hartrees.

[a] J. B. Anderson, C. A. Traynor, and B. M. Boghosian, Quantum chemistry by random walk: Exact treatment of many-electron systems, *J. Chem. Phys.* **95**, 7418 (1991).

86

V. BUCH

Treatment of rigid bodies by diffusion Monte Carlo:
Application to the para-H_2–H_2O and ortho-H_2–H_2O
clusters

J. Chem. Phys. **97**, 726–729 (1992)

This paper describes the application of diffusion QMC to the problem of motions of molecules in small clusters in which some or all of the molecules can be treated as rigid bodies. For the loosely bound ground state of H_2–H_2O clusters the high-frequency intramolecular vibrations of H atoms in H_2 and in H_2O are decoupled from the lower-frequency intermolecular vibrations and both H_2 and H_2O can be considered rigid bodies. A potential energy surface based on rigid bodies was obtained as an approximate fit to data from ab initio calculations. It has three minima for H_2 in the plane of the H_2O molecule and two out of the plane.

The diffusion calculations were carried out in angular coordinates for the rigid rotor motions and cartesian coordinates for the H_2–H_2O translational motions. In some cases the H_2O molecule was completely fixed. For para-H_2 and para-H_2O the ground state wavefunctions are fully symmetric and without nodes so that no restrictions were necessary. For ortho-H_2 and para-H_2O a cancellation procedure for walkers with opposite H_2 rotation vectors was required to impose the antisymmetry condition. In this case the rotational state of free H_2 is triply degenerate and the degeneracy was observed to remain in the presence of H_2O with distributions of walkers drifting among the states. For para-H_2 and a fixed H_2O molecule the favored position of H_2 is near one of the H atoms of water with nearly free rotation around the O-H axis. For ortho-H_2 with preferred orientations in space the corresponding position is less populated and the distribution is shifted to positions near the O atom. These results are consistent with interpretations of infrared spectra for the H_2–H_2O complex in an argon matrix. The calculations for allowed rotations of both H_2 and H_2O showed a roughly spherical distribution of H_2 position around the H_2O.

87

D. L. DIEDRICH & J. B. ANDERSON

An accurate quantum Monte Carlo calculation of the
barrier height for the reaction H + H$_2$ → H$_2$ + H

Science **258**, 786–788 (1992)

A list of fundamental reactions usually begins with the reaction
H + H$_2$ → H$_2$ + H, the simplest of neutral chemical reactions and
the subject of more than 60 years of theoretical studies at the
time this paper was published. The potential energy surface and
especially the barrier to reaction are essential to any treatment.
Earlier theoretical work had led to predictions of a barrier height
of 9.5 to 9.8 kcal/mol. The QMC calculations reported in this paper
reduced the uncertainty by a factor of about ten and specified the
barrier height as 9.61 ± 0.01 kcal/mol.

The calculations were "exact" QMC calculations based on the
Green's function method with partial cancellations of positive and
negative walkers to produce a stable wavefunction with a correct
node location. A trial wavefunction was used for importance sampling to aid in determining energies without affecting the node
location. Time-step error was avoided with use of the Green's
function approach for the moves of walkers. The complete procedure was exact in that, except for the easily estimated statistical
or sampling error, it required no mathematical or physical assump-
tions beyond those of the Schrödinger equation. No interpolation
or extrapolation was required.

The calculations required a few months on a workstation computer at the time. A year earlier Clementi et al.[a] had speculated
that an analytic variational calculation of similar accuracy "would
call for somewhere around 3500 years" on a similar computer.

[a]E. Clementi et al., Selected topics in ab initio computational chemistry
in both very small and very large chemical systems, *Chem. Rev.* **91**, 679 (1991).

88

P. BALLONE, C. J. UMRIGAR, & P. DELALY

Energies, densities, and pair correlation functions of jellium spheres by the variational Monte Carlo method

Phys. Rev. B **45**, 6293–6296 (1992)

In the jellium model for clusters of atoms the charge of the nuclei and of appropriate inner-shell electrons is replaced by a net positive uniform charge distribution. This model has been found especially useful for spherical clusters of alkali metal atoms with explicit treatment of electronic structure limited to valence electrons.

In this paper the authors have reported studies of clusters of up to 20 atoms, equivalent to spheres with a distributed charge of +20 along with 20 free electrons. The trial function for variational QMC was constructed in the usual way with single determinants of orbitals generated in density functional (LSDA) calculations multiplied by optimized Jastrow functions. The quality of these functions was tested in fixed-node diffusion calculations indicating that 90% of the fixed-node correlation energy was recovered in the variational calculations.

Results for a series of neutral and ionized clusters were compared with those of density functional calculations in terms of total, exchange, and correlation energies as well as pair correlation functions and electron densities. In general, total and spin densities were found in excellent agreement for the two types of calculations. Energies were found in fairly good agreement, but exchange and correlation energies differed by as much as 6% and 30%, respectively. The paper clearly shows the several advantages of QMC for a system different from that of the typical atomic or molecular problem.

89

R. F. BISHOP, E. BUENDIA, M. F. FLYNN, & R. GUARDIOLA

Diffusion Monte Carlo determination of the binding
energy of the ^{4}He nucleus for model Wigner potentials

J. Phys. G: Nucl. Part. Phys. **18**, L21–L27 (1992)

The diffusion QMC method is especially well suited for a number of problems in areas far from the electronic structure theory of atomic and molecular systems. This paper describes a set of problems in the binding of nucleons for which a wide variety of other methods (mostly variational, including independent pair correlations, configuration interaction, coupled cluster, coupled rearrangement, and many others) have given some accurate results, but for which the DQMC method gives essentially exact results. The example here is the two-proton, two-neutron ^{4}He nucleus with four different model Wigner potentials that omit some of the basic features of the problem but may be considered "quasi-realistic." The four model potentials, known as V7, B1, S3, and MTV, are of a form such that the ground state is nodeless, thus providing ideal cases for DQMC.

Both variational and diffusion QMC calculations were carried out for each of the model potentials using a trial wavefunction consisting of a product of Gaussian functions, one for each nucleon, multiplied by a nucleon-nucleon Jastrow function, all optimized to minimize the VQMC energy (not the variance). The VQMC energies so obtained were in broad agreement with those of other methods, but the DQMC energies were lower and for each of the four interactions may be considered the most accurate available. As noted in the paper, these provide "an extremely accurate set of reference numbers against which other approximate techniques . . . may be tested and refined."

90

P. BELOHOREC, S. M. ROTHSTEIN, & J. VRBIK

Infinitesimal differential diffusion quantum Monte Carlo study of CuH spectroscopic constants

J. Chem. Phys. **98**, 6401–6405 (1993)

At the time this paper was written the heaviest atom that had been treated in all-electron QMC calculations was the fluorine atom. This study moved all the way to the copper atom in calculations for the diatomic molecule CuH. The technique used was a differential method introduced earlier[a] by the authors. In this study the vibration-rotation spectroscopic constants for the molecule were determined from the first four derivatives of the total energy with respect to the internuclear distance evaluated at the equilibrium distance. With the aid of the Hellmann-Feynman theorem these were determined from the first four derivatives of the potential energy. The problem of differing time scales for the large-Z atom was solved by using a split-τ method in which the electrons of Cu were divided into four groups according to their distances from the nucleus. Those in each group were moved with different time steps in the series $0.016\,\tau$, $0.040\,\tau$, $0.120\,\tau$, and τ. The question of how to treat branching in diffusion QMC was avoided by eliminating branching altogether so that the treatment was equivalent to variational QMC. The parameter τ was varied to allow extrapolation to zero time-step size. The trial wavefunction for importance sampling was a minimal SCF function composed of Slater orbitals optimized at the experimental equilibrium internuclear distance to obtain a minimum variance in local energy. The calculations produced the vibration-rotation spectroscopic constants ω_e, $\omega_e x_e$, and α_e and the dipole moment. These were found in reasonable agreement with experimental values, not as good as those of the best CI calculations of the time but quite good for the very simple trial function used. The use of better trial functions and investigation of means for combining branching with the split-τ method were suggested.

[a] J. Vrbik, D. R. Legare, and S. M. Rothstein, Infinitesimal differential diffusion quantum Monte Carlo: Diatomic molecular properties, *J. Chem. Phys.* **92**, 1221 (1990).

91

D. M. SCHRADER, T. YOSHIDA, & K. IGUCHI

Binding energies of positronium fluoride and positronium bromide by the model potential quantum Monte Carlo method

J. Chem. Phys. **98**, 7185–7190 (1993)

This paper and a companion paper[a] describe three pseudopotential QMC calculations for the positronium halides PsF, PsCl, and PsBr with accuracies and reliabilities not previously available. These three compounds have a fleeting lifetime sufficiently long that time-independent nonrelativistic quantum mechanics should give good estimates of their energies but too short to allow direct measurements of those energies. Earlier estimates of several types and calculations at the Hartree-Fock level had given a wide spread in predictions of the binding energies of positronium to the halide ions, mostly in the range of 1 to 3 eV for the three molecules.

The calculations were fixed-node diffusion QMC calculations for the eight valence electrons and the positron with a model potential expression replacing the core electrons and halide nucleus. The model or pseudopotential used was a simplified version without valence-core correlation and with several lesser terms eliminated in order to make it a local pseudopotential. This model[b] had been used previously in successful predictions of the electron affinity of the Cl atom. An importance sampling trial wavefunction was constructed of Hartree-Fock orbitals for the halide ion and a $2s$ orbital for the positron, along with an electron-positron term but no electron-electron correlation terms. Checks of the consistency of the calculations with independent estimates of the correlation energies were entirely satisfactory. The results, in terms of binding energies, were obtained in the range of 1–2 eV (with uncertainties of 0.1–0.2 eV) in an order consistent with competitive formation experiments.

[a]D. M. Schrader, T. Yoshida, and K. Iguchi, Binding energy of positronium chloride: A quantum Monte Carlo calculation, *Phys. Rev. Lett.* **68**, 3281 (1992).

[b]T. Yoshida, Y. Mizushima, and K. Iguchi, Electron affinity of Cl: A model potential — quantum Monte Carlo study, *J. Chem. Phys.* **89**, 5815 (1988).

92

J. B. ANDERSON, C. A. TRAYNOR, & B. M. BOGHOSIAN

An exact quantum Monte Carlo calculation of the helium-helium intermolecular potential

J. Chem. Phys. **99**, 345–351 (1993)

The first theoretical estimate based on quantum mechanics for the minimum in the potential energy of interaction of two helium atoms was published by Slater in 1928, 65 years before this QMC paper. In the intervening years some 50 other papers on the subject were published. None was very successful in predicting the well depth until counterpoise corrections were introduced to reduce the error caused by "basis set superposition," a term first used for the helium-helium system. By 1991 the best of the counterpoise-corrected CI calculations gave well depths of about 10.93 K. These were obtained from energy differences in calculations for separated and for interacting atoms having errors of about 1200 K in the total energy. The QMC calculations were carried out using the "exact" cancellation method[a] applicable to few-electron systems for which the node problem is solved exactly to obtain energies without physical or mathematical approximations beyond those of the Schrödinger equation. The remaining error, the statistical error in the total energy, was only 0.10 K. That was 1/12000 the error in the total energy for the CI calculations. At the equilibrium separation of 5.6 bohr the well depth was found to be 11.01 ± 0.10 K. The complete potential energy curve was entirely consistent, within the probable errors, with experimentally derived potential energy curves and its error bars straddled the best of available counterpoise-corrected calculations. It was pointed out that more accurate QMC values could be obtained with additional calculations using the same computer program. (That was done in 2004 to yield a well depth of 10.998 ± 0.005 K.)[b]

[a] J. B. Anderson, C. A. Traynor, and B. M. Boghosian, Quantum chemistry by random walk: Exact treatment of many-electron systems, *J. Chem. Phys.* **95**, 7418 (1991).

[b] J. B. Anderson, Comment on an exact quantum Monte Carlo calculation of the helium-helium intermolecular potential, *J. Chem. Phys.* **120**, 9886 (2004).

93

A. BHATTACHARYA & J. B. ANDERSON

Exact quantum Monte Carlo calculation of the H-He
interaction potential

Phys. Rev. A **49**, 2441–2445 (1994)

This is the third in a series of papers describing "exact" QMC
calculations for few-electron systems with nodes in their wavefunctions. The method used is the exact cancellation Green's function
scheme[a] which develops the node structure and is "exact" in that
it requires no approximations, no interpolations, and no extrapolations. The treatment used is essentially the same as that for
earlier calculations for He ^3P, H_2 $^3\Sigma_u^+$, H-H-H, and He-He. The
trial function for importance sampling was based on a Hylleraas
function for the He atom and a Slater function for the H atom.

Energies were calculated for internuclear separations within the
range of 5.0 to 15.0 bohr. The minimum was found at $-3.403\,746\pm$
$0.000\,002$ hartrees for a separation of 7.0 bohr, corresponding to a
well depth of 6.8 ± 0.5 K. Total energies were approximately 12000
K below those of the lowest-energy variational calculations available. The interaction energy was at least 3 K more accurate than
prior estimates by any other method. The calculated potential energy curve was consistent with low-energy scattering cross-section
measurements for H with He.

[a] J. B. Anderson, C. A. Traynor, and B. M. Boghosian, Quantum chemistry
by random walk: Exact treatment of many-electron systems, *J. Chem. Phys.*
95, 7418 (1991).

94

L. MITÁŠ

Quantum Monte Carlo calculation of the Fe atom

Phys. Rev. A **49**, 4411–4414 (1994)

A variety of earlier calculations for the third-row transition element iron using argon-core pseudopotentials had failed to give good agreement with experimental measurements of the ionization potential, the electron affinity, and excitation energies. This paper reports QMC calculations which give excellent agreement and which are likely to be the most accurate of any type of calculation for the Fe atom in the 1990s. For these the author used neon-core pseudopotentials (leaving 16 electrons, including $3s$ and $3p$ electrons, in the valence space) for both variational and fixed-node diffusion QMC calculations. The pseudopotential used was an ab initio pseudopotential tested against all-electron results in limited configuration interaction calculations. Relativistic effects were included. The complete Hamiltonian was modified slightly to facilitate treatment of the nonlocal part of the pseudopotential expression. The trial functions for importance sampling and specifying nodes were linear combinations of Slater determinants along with Jastrow functions.

Six low-lying neutral states were examined in addition to the ground states of the anion and cation. Recovery of valence correlation energies was 0.6 to 0.8 hartrees for all of these, typically 0.1 to 0.2 hartrees greater that of coupled cluster and configuration interaction calculations. Most impressive is the agreement with experimental measurements: ionization potential, DQMC 7.67(6) eV vs. expt. 7.87 eV; $^5D \rightarrow {}^3F$ transition, DQMC 4.24(9) eV vs. expt. 4.07 eV; $^5D \rightarrow {}^5F$ transition, DQMC 0.84(6) eV vs. expt. 0.87 eV; electron affinity, DQMC -0.03(9) eV vs. expt. 0.15 eV.

95

A. B. FINNILA, M. A. GOMEZ, C. SEBENIK, C. STENSON, & J. D. DOLL

Quantum annealing: A new method for minimizing multidimensional functions

Chem. Phys. Lett. **219**, 343–348 (1994)

Quantum Monte Carlo has found a new application far afield from the solution of the Schrödinger equation for many-body problems. In this paper the authors report a method for solving multidimensional optimization problems that is superior, in many cases, to other methods such as conjugate gradient methods, the simplex method, direction-set methods, genetic methods, and classical simulated annealing. The new approach, aptly titled "quantum annealing," is closely related to classical annealing, in which a physical system such as a metal cluster is slowly cooled and finds its lowest-energy configuration as the temperature drops to zero. In classical annealing the system follows classical-mechanical dynamics as the temperature drops. In quantum annealing a collection of configurations evolves in the the same way as walkers undergoing diffusion and multiplication in diffusion QMC with the added feature of a slowly dropping value for Planck's constant h. Classical annealing allows the escape from local minima with the aid of thermal fluctuations. Quantum annealing allows such escape by delocalization and tunneling. Since a QMC walker can multiply in number on entering a low-energy region, a single walker can populate an entire region.

The paper reports several example applications beginning with a one-dimensional double-well problem and ending with a 19-atom Lennard-Jones cluster having about 10^5 local minima. In each of the cases described quantum annealing was successful in finding the global minimum.

96

M. LEWERENZ & R. O. WATTS

Quantum Monte Carlo simulation of molecular vibrations: Application to formaldehyde

Mol. Phys. **81**, 1075–1091 (1994)

The formaldehyde molecule is an often used test system for methods of predicting vibrational levels of polyatomic molecules. It was the first four-atom molecule to be treated in analytic variational calculations and it was the subject of early QMC variational calculations by Bernu et al.[a] It has also been the subject of several perturbational studies. The study reported in this paper has taken advantage of a preexisting accurate global potential energy surface, providing good descriptions of the several local minima, for fully QMC calculations using the fixed-node diffusion method.

The fixed nodes for several of the excited states treated could be specified exactly from symmetry considerations. For the others approximate node locations were used. Importance-sampling trial wavefunctions were assembled from diatomic and triatomic subunits in "modular" constructions. Calculations for a total of eight different vibrational states gave lower energies than variational calculations for the same surface but good agreement in frequencies except for two combination states. Agreement with experiment in terms of vibrational frequencies was good, especially for the lowest states of a given symmetry.

[a]B. Bernu, D. M. Ceperley, and W. A. Lester, Jr., The calculation of excited states with quantum Monte Carlo. II. Vibrational excited states, *J. Chem. Phys.* **93**, 552 (1990).

97

L. MITÁŠ & R. M. MARTIN

Quantum Monte Carlo of nitrogen: Atom, dimer, atomic and molecular solids

Phys. Rev. Lett. **72**, 2438–2441 (1994)

This paper reports the first electronic structure calculations of any type to treat nitrogen solids with very high accuracies — accuracies similar to those obtained for atomic and molecular nitrogen. It also reports the first estimate of a band gap for a solid using QMC methods. The calculations were carried out as valence-only calculations with ab initio nonlocal pseudopotentials using diffusion QMC with importance sampling. For the atomic and molecular solids, compressed $I2_13$ and $Pa3$ crystals, cubic structures with eight atoms in a simulation cell and periodic boundary conditions were used. The trial functions for importance sampling were composed of Slater determinants from Gaussian92, Gamess, and Crystal92 calculations along with correlation terms optimized in variational QMC calculations.

For the atom three different forms for the pseudopotential were found to give identical correlation energies. The pseudopotential of Stevens, Basch, and Krauss[a] was subsequently used for all systems. For the dimer the DQMC calculations gave a binding energy within 2% of the experimental binding energy . . . a result "comparable with the accuracy of the most extensive quantum chemistry calculations." For the two compressed solids the binding energies were found to be 3.3(1) and 3.4(1) eV, values much higher than those of Hartree-Fock calculations, 0.01 and 0.25 eV, and lower than those of LDA density functional calculations at 4.45 and 4.75 eV. No experimental energies were available. The energy gap for $\Gamma \rightarrow H$ was determined for the $I2_13$ structure with added calculations for excited states, each the lowest state of a symmetry class different from that of the ground state. The results lead to an estimated band gap of 8.5(4) eV. This represents a significant improvement over the HF and LDA estimates of 18.0 and 6.1 eV, respectively, that err in the same way as found in earlier tests.

[a]W. J. Stevens, H. Basch, and M. Krauss, Compact effective potentials and efficient shared-exponent basis sets for the first- and second-row atoms, *J. Chem. Phys.* **81**, 6026 (1984).

98

F. BOLTON

The effect of an uneven potential on the few-electron
spectrum in a quantum dot

Physica B **212**, 218–223 (1995)

Quantum Monte Carlo methods devised for atomic and molecular systems may be applied without much change to investigate
the physics of quantum dots. The simplest of few-electron quantum dots is the idealized case with two electrons confined in a
two-dimensional well by a parabolic potential, which is a good approximation to the confining potential of a uniform background of
positive charge. With more electrons the node problem appears,
and with an applied magnetic field the Hamiltonian gains more
terms and the wavefunction may be treated as complex. In this
study, the effects of adding ripples to the parabolic potential, such
as might be found in realistic confining potentials, were investigated
for one-, two-, and three-electron systems.

Diffusion QMC was used with the equivalent of fixed nodes,
which were specified as phase factors in the trial functions in what
is termed the fixed-phase approximation. These can be specified for
low-lying singlet and triplet states. With the magnetic field varied,
the ground state may shift to a state of different total spin. The addition spectrum, corresponding to the energy changes on addition
of an electron to the system, and its changes with magnetic field
were calculated for conditions matching experiments with GaAs.
The calculated and observed spectra were nearly the same, but
some differences were found. These were explained in part by additional calculations with realistic ripples added to the parabolic
potentials.

99

B. CHEN & J. B. ANDERSON

Improved quantum Monte Carlo calculation of the
ground-state energy of the hydrogen molecule

J. Chem. Phys. **102**, 2802–2805 (1995)

Since the ground state of the hydrogen molecule is nodeless, it is easily treated by "exact" QMC cancellation methods in a full 12-dimensional, four-particle coordinate space to yield the ground-state energy without the use of the Born-Oppenheimer approximation. Following the discovery of quantum mechanics, the hydrogen molecule has been the frequent target of theoretical predictions of increasing accuracy, in parallel with experimental measurements of its ionization potential and its dissociation energy, also with increasing accuracy. The calculations reported in this paper took advantage of importance sampling and efficient cancellation to yield an accuracy of better than 1 microhartree in the total energy, with the result of $-1.164\ 0237 \pm 0.000\ 0009$ hartrees, a value slightly less accurate than analytic variational calculations at the time. Expressed as the dissociation energy and corrected for relativistic and radiative effects, the QMC result in 36117.84 ± 0.20 cm^{-1}, which may be compared with the experimental value[a] of 36118.11 cm^{-1} determined in 1992.

[a] A. Balakrishnan, V. Smith, and B. P. Stoicheff, Dissociation energy of the hydrogen molecule, *Phys. Rev. Lett.* **68**, 2149 (1992).

100

S. D. KENNY, G. RAJAGOPAL, & R. J. NEEDS

Relativistic corrections to atomic energies from
quantum Monte Carlo calculations

Phys. Rev. A **51**, 1898–1904 (1995)

This paper describes the first calculations incorporating electron correlation effects in the prediction of relativistic corrections for atoms as large as Ne. The importance of correlation in relativistic corrections had been previously shown for two-electron atoms in conventional calculations and VQMC calculations[a] for the LiH molecule. This study extends such treatments to ten electrons in VQMC and DQMC calculations. The calculations are perturbational and are based on an effective Hamiltonian correct to order $1/c^2$, where c is the speed of light. Both variational and diffusion QMC calculations were performed for the 2-electron systems He, Be^{2+}, and Ne^{8+}, for the 4-electron atom Be, and for the 10-electron atom Ne. For each species, the energy shifts due to each of the several terms in the Hamiltonian could be evaluated as the ψ_T^2 integral in VQMC or the $\psi\psi_T$ integral in DQMC. The $\psi\psi$ integral values were estimated as $(2\psi\psi_T - \psi_T^2)$ integrals. The total relativistic effects ranged from -0.000082 hartrees for He to -0.135 hartrees for Ne. For He, Be^{2+}, and Ne^{8+} the calculated shifts are in good agreement with nearly exact results of earlier calculations by other methods, and VQMC and DQMC results were very close to each other. For Be, the shift is in good agreement with experiment. For Ne, the shift is not in good agreement, but the experimental value is more uncertain. The effects of electron correlation were found to reduce the absolute values of the relativistic corrections for all species. The specific relativistic correction for Ne changed from -0.145 hartrees without correlation to -0.141 with correlation. The authors noted that QMC offers a simple and accurate means for evaluating perturbative relativistic corrections for atoms of low atomic number.

[a] J. Vrbik, M. F. DePasquale, and S. M. Rothstein, Estimating the relativistic energy by diffusion quantum Monte Carlo, *J. Chem. Phys.* **88**, 3784 (1988).

101

C. HUISZOON & M. CAFFAREL

A quantum Monte Carlo perturbational study of the
He-He interaction

J. Chem. Phys. **104**, 4621–4631 (1996)

The well depth for interaction of two helium atoms at their equilibrium internuclear distance is less than $1/10^5$ the total electronic energy of the pair. Thus, a perturbational approach is a natural way to attack the problem of predictions of the interaction energy, and several such attempts have been made. The work described in this paper is perturbational, but it differs from previous work in that it uses a QMC method, specifically simple diffusion QMC, to determine the values of the perturbation terms required. The interaction energy for the two helium atoms was expressed by a sum of Rayleigh-Schrödinger (RS) terms and exchange terms as:

$$\Delta E_{\text{int}} = E_{RS}^{(1)} + E_{\text{exch}}^{(1)} + E_{RS}^{(2)} + E_{\text{exch}}^{(2)} + E_{RS}^{(3)}.$$

Caffarel and Hess[a] showed earlier that the RS terms could be evaluated in terms of multitime integrals of autocorrelation functions along the trajectories of a diffusion process related to the unperturbed Hamiltonian and that these integrals could be determined by solution of the Fokker-Planck diffusion equation exactly, as in simple diffusion QMC calculations. The exchange terms could be similarly evaluated. The calculations were carried out for several internuclear distances and, within their relatively large statistical errors, gave agreement with "exact" values[b] given by other QMC methods.

[a] M. Caffarel and O. Hess, Quantum Monte Carlo perturbation calculations of interaction energies, *Phys. Rev. A* **43**, 2139 (1991).

[b] J. B. Anderson, C. A. Traynor, and B. M. Boghosian, An exact quantum Monte Carlo calculation of the helium-helium interaction potential, *J. Chem. Phys.* **99**, 345 (1993).

102

R. E. TUZUN, D. W. NOID, & B. G. SUMPTER

An internal coordinate quantum Monte Carlo method for calculating vibrational ground state energies and wave functions of large molecules: A quantum geometric statement function approach

J. Chem. Phys. **105**, 5494–5502 (1996)

The problem described in this paper is the calculation of the ground-state energy for a 100-atom model polyethylene chain. The chain is represented by 100 mass points, connected in a series and interacting sequentially according to bond stretch terms, one-angle bond terms, and two-angle torsion terms. The lowest-energy configuration is a planar zigzag arrangement. The authors had earlier devised an internal coordinate system to facilitate classical dynamics simulations of such chains, and they used these again to great advantage in diffusion QMC calculations for this problem. In preliminary investigations, it was found that importance sampling was essential. The trial function was expressed as a product of Gaussian terms in each of the internal coordinates. With these the required first and second derivatives, as well as the drift terms, were reduced to very simple, very compact expressions. A series of calculations at different time-step sizes was extrapolated to obtain the ground-state energy. The value of 211 kcal/mol obtained in this way was about 7% below that obtained in a normal coordinate analysis. The application of diffusion QMC to the efficient solution of a long-standing problem is nicely demonstrated in this work.

103

B. CHEN, M. A. GOMEZ, M. SEHL, J. D. DOLL, & D. L. FREEMAN

Theoretical studies of the structure and dynamics of metal/hydrogen systems: Diffusion and path integral Monte Carlo investigations of nickel and palladium clusters

J. Chem. Phys. **105**, 9686–9694 (1996)

In this study, the effect of adsorption of a single hydrogen atom upon the properties of Ni and Pd clusters of up to ten atoms was investigated. The required potential energies for the interactions of the atoms within the clusters were determined with the embedded-atom method, a semi-empirical method with parameters fit to data for bulk systems, giving reasonable agreement in comparisons with experimental observations. With the potential energies then available in convenient form, the properties of the clusters could be investigated to determine minimum-energy structures with and without zero-point energies, preferred hydrogen binding sites, hydrogen vibration frequencies, and related properties as a function of cluster size, using classical and quantum Monte Carlo methods.

For the smaller clusters, the lowest-energy structures occur with the hydrogen atom on the outside of the clusters, but for the larger, inclusion of the hydrogen atom within the cluster is favored. Zero-point vibrations were found to have no significant effects on the favored structures. Classical harmonic vibration analyses for the entire clusters gave about the same zero-point energies as diffusion QMC calculations. For the ground states, the QMC calculations showed the hydrogen atoms to be localized. For some clusters, Pd_7H and $Pd_{10}H$, quantum effects as indicated by diffusion QMC were important in determining the favored structure. The studies were carried further with finite-temperature energy averages determined with the Fourier path integral method. The variation of energy with temperature clearly shows the quantum behavior of the smaller clusters at low temperatures and the shift toward classical behavior with increasing cluster size.

104

D. F. R. BROWN, J. K. GREGORY, & D. C. CLARY

A method to calculate vibrational frequency shifts in heteroclusters: Application to N_2^+-He_n

J. Chem. Soc., Faraday Trans. **92**, 11–15 (1996)

Ionic clusters are important in a wide variety of chemical systems, especially in those involving ion-molecule reactions in the atmosphere and in biological systems, but they received relatively little attention until the mid-1990s. The N_2^+-He_n system is one system subjected to spectroscopic studies and theoretical examination at that time. In this paper, the authors reported diffusion QMC calculations for the vibrations within these clusters and the development of a new method for predicting the N-N vibrational frequency shifts induced by the presence of the loosely attached helium cluster He_n of varying n. Minimum-energy and vibrationally averaged structures were also reported.

The QMC calculations were performed for motion on a pairwise-additive potential energy surface fitted to ab initio electronic structure calculations. The weak coupling of the high-frequency motion of the N-N vibration with the lower frequency motions of the He atoms in the cluster allowed a separation of these motions and the treatment of the complete wavefunctions as a product of separate wavefunctions for N_2^+ and He_n. Calculations were carried out for clusters with up to 16 helium atoms. For each of these clusters, a number of calculations were made with the N-N distance frozen. The simulations produced an adiabatic potential energy curve for N-N motion, from which vibration frequencies could be determined. The frequency shift was found to vary linearly with the number of helium atoms n. No experimental data were available for direct comparisons. The authors note the general applicability of the method to similar problems, including those of much higher complexity.

105

S. D. KENNY, G. RAJAGOPAL, R. J. NEEDS, W.-K. LEUNG, M. J. GODFREY, A. J. WILLIAMSON, & W. M. C. FOULKES

Quantum Monte Carlo calculations of the energy of the relativistic homogeneous electron gas

Phys. Rev. Lett. **77**, 1099–1102 (1996)

Several earlier QMC calculations[a,b] had been successful in predicting the energy of the homogeneous electron gas. This paper extended the earlier work to include relativistic effects with use of first-order perturbation theory for a wide range of densities. The required perturbation expression was derived in the 1950s. It consists of four terms known as the mass-velocity, Darwin, contact, and retardation terms. The small perturbative correction to the nonrelativistic energy is given by the sum of the expectation values of the individual terms The wavefunctions were obtained from variational and fixed-node diffusion calculations.

The system was treated as a face-centered cubic simulation cell with periodic boundary conditions, containing up to 338 electrons for the variational and 178 electrons for the diffusion calculations. The trial functions ψ_T were composed of Slater determinants of plane waves, along with Jastrow factors. The required integrals based on ψ_T^2 were accumulated with Metropolis sampling, and those based on $\psi\psi_T$ were accumulated with importance-sampling diffusion simulations. Values based on ψ^2 were obtained by extrapolation The mass-velocity term was found to dominate at high densities and to agree with Hartree-Fock results. At lower densities the mass-velocity term from QMC was a factor of three larger than for HF, but the retardation term was dominant and much larger than predicted by HF. Overall, the results show clearly the importance of correlation, even for relativistic effects.

[a] D. M. Ceperley and B. J. Alder, Ground state of the electron gas by a stochastic method, *Phys. Rev. Lett.* **45**, 566 (1980).

[b] G. Ortiz and P. Ballone, Correlation energy, structure factor, radial-distribution function, and momentum distribution of the spin-polarized electron gas, *Phys. Rev. B* **50**, 1391 (1994); ibid. **56**, 9970 (1997).

106

K. LIU, M. G. BROWN, C. CARTER, R. J. SAYKALLY, J. K. GREGORY, & D. C. CLARY

Characterization of a cage form of the water hexamer

Nature **381**, 501–503 (1996)

This paper received attention in the news of 1996 for providing, at least in part, an answer to the question, "How many molecules of H_2O are required to make a drop of water?" It reports an experimental investigation of the spectroscopy of the water hexamer and companion QMC calculations indicating the structure of the hexamer is cage-like. Since the trimer, tetramer, and pentamer are most likely cyclic in structure, the three-dimensional cage-like cluster of the hexamer would qualify it as the smallest possible drop.

The structure of the hexamer was investigated with terahertz laser vibration-rotation spectroscopy, which established the cluster size and vibration-rotation splittings consistent with a cage structure held together with eight hydrogen bonds, in agreement with the QMC calculations. These were carried out in diffusion QMC to determine the vibrationally averaged ground-state rotational constants for the five lowest-energy potential energy minima for the hexamer. Since these involved 30 strongly coupled degrees of freedom, diffusion QMC was the method of choice. The results showed that although a prism structure has the lowest potential energy minimum, the cage structure is slightly more stable due to a lower zero-point energy. Vibrational averaging has a large effect on the moments of inertia (and therefore the rotational constants) for the cage, and it brings them into good agreement with the experimental values for these small "drops" of water.

M. LEWERENZ

Structure and energetics of small helium clusters:
Quantum simulations using a recent perturbational pair
potential

J. Chem. Phys. **106**, 4596–4603 (1997)

Small helium clusters He_n $(n = 2, 3, \ldots, 10)$ in molecular beams produced by supersonic expansion of gaseous helium have been observed in several different experiments. Among these clusters the dimer, which has a single very slightly bound state, and the trimer, which could have many bound states, have especially unusual properties which QMC calculations can explain in detail. In this paper, diffusion QMC calculations were used to determine ground state properties for clusters $(n = 2, \ldots, 10)$ using two accurate, but slightly different pairwise-additive potential energy expressions, a perturbation-based potential by Tang et al.[a] and an empirical potential (HFD-B) proposed by Aziz et al.[b] Trial functions for importance sampling were based on those determined earlier by Rick et al.[c] Expectation values and probability distributions were determined by the descendent-weighting technique to give results independent of the trial functions. As expected, a single bound state of the dimer was found, and its average internuclear distance was predicted to be 51.5 Å. Energies for clusters with $n = 2, 7$ with the HFD-B potential were in good agreement with those of earlier calculations. Comparisons of the results for the two different potential energy expressions showed energies differing by a few percent. Tests of possible nonadditive interactions showed these to be far less important in their effects on predicted energies.

[a] K. T. Tang, J. P. Toennies, and C. L. Yiu, Accurate analytical He-He van der Waals potential based on perturbation theory, *Phys. Rev. Lett.* **74**, 1546 (1995).
[b] R. A. Aziz, F. R. W. McCourt, and C. C. K. Wong, A new determination of the ground-state interatomic potential for He_2, *Mol. Phys.* **61**, 1487 (1987).
[c] S. W. Rick, D. L. Lynch, and J. D. Doll, A variational Monte Carlo study of argon, neon, and helium clusters, *J. Chem. Phys.* **95**, 3506 (1991).

108

R. N. BARNETT, Z. SUN, &
W. A. LESTER, JR.

Fixed-sample optimization in quantum Monte Carlo
using a probability density function

Chem. Phys. Lett. **273**, 321–328 (1997)

The optimization of trial wavefunctions ψ_T for use in QMC calculations can be accomplished with a fixed set of configurations selected with probabilities proportional to ψ_T^2, often using Metropolis sampling and occasionally using other methods. By varying the coefficients of ψ_T, one can minimize the variance in local energies for the original set of configurations. This is effective except when coefficients affecting the node locations are varied and occasional very high weighting factors $\psi_T^2(\text{new})/\psi_T^2(\text{original})$ can give large fluctuations in the variance. In this paper, the authors propose the use of probability density functions ρ to produce nodeless distributions of points replacing the original ψ_T^2 distributions. This has two advantages: it avoids the unbounded weighting problem allowing nodes to be varied, and it facilitates sampling without a Metropolis walk and its associated serial correlation. To be effective, i.e., for high efficiency, ρ must be carefully selected, it must at least resemble ψ_T^2, and its distributions must be easily sampled. Tests were performed for the case of CH with optimization of all parameters of a two-determinant-and-Jastrow trial function. These included basis set exponents, MO coefficients, CI coefficient, and Schmidt-Moskowitz Jastrow parameters. The resulting variance in local energy was 0.19 h^2, versus 0.25 h^2 for optimization of the Jastrow coefficients only. More interesting is the improvement in diffusion QMC energy for the complete optimization including node movement. The energy was lowered by 0.004 hartree, or 2.5 kcal/mol, with recovery of correlation energy increased from 92.8 to 94.8%.

109

H.-J. FLAD, M. DOLG, & A. SHUKLA

Spin-orbit coupling in variational quantum Monte Carlo calculations

Phys. Rev. A **55**, 4183–4195 (1997)

Although several types of spin interactions had been included in earlier QMC studies in nuclear physics, this is the first paper to report QMC calculations including the spin-orbit effect for atomic systems. The problem treated is that of predicting the effect of valence electron correlation on the energies of eigenstates involving the $6s^2\,6p^2$ valence electrons of the isoelectronic species Pb, Bi^+, and Po^{2+}. Experimental measurements of the fine-structure splittings provided a challenging target for the authors. It was necessary to simplify the problem with a series of approximations, beginning with a pseudopotential to eliminate the 78 electrons of the Pt core. A semi-local spin-orbit-averaged relativistic pseudopotential was used for this purpose. With total spin and spatial angular momentum variable, the trial wavefunctions were expressed as linear combinations of LS-coupled wavefunctions, similar in form to functions without spin-orbit coupling, but very much more complicated in detail. The calculations were carried out in variational QMC with a generalized Metropolis algorithm, in effect diffusion with drift, but without multiplication. The results were expressed as excitation energies for the $6s^2\,6p^2$ $^3P_1, ^3P_2, ^1D_2$, and $^1S_0$ electronic states of Pb, Bi^+, and Po^{2+} with respect to the $^3P_0$ ground state. The values were typically accurate within a few percent and comparable to those of equivalent all-electron and pseudopotential calculations by multireference configuration interaction methods.

110

R. Q. HOOD, M. Y. CHOU,
A. J. WILLIAMSON, G. RAJAGOPAL,
R. J. NEEDS, & W. M. C. FOULKES

Quantum Monte Carlo investigation of exchange and correlation in silicon

Phys. Rev. Lett. **78**, 3350–3353 (1997)

QMC calculations can be used to calculate wavefunctions, electron densities, and a wide variety of functions and integrals, all of which may in turn be useful in improving other types of electronic structure calculations. In this variational QMC study of diamond-structure silicon, the quantities of central importance in Kohn-Sham density functional theory were calculated and compared to those from local-density and average-density approximations. In this case, the quantities were the coupling-constant-integrated pair correlation function, the exchange-correlation hole, and the exchange-correlation energy density. All are related to the total exchange-correlation energy.

The calculations were carried out for $3 \times 3 \times 3$ fcc unit cells containing 54 nuclei with pseudopotentials and 216 valence electrons. The VQMC wavefunctions were optimized Slater-Jastrow functions giving 85% of the fixed-node correlation energy. Several different techniques were used in sampling to accumulate the diamond quantities, which were illustrated in full-color contour plots showing variations with position in the lattice. Differences between VQMC and local density approximation (LDA) results were similarly illustrated, and the largest errors in the LDA exchange-correlation energy densities were found in the bonding regions and around the pseudoatoms. As pointed out by the authors, such calculations should aid in developing better functionals by providing much more data than the total energy and the bulk moduli previously available for comparison.

111

V. N. STAROVEROV, P. LANGFELDER, & S. M. ROTHSTEIN

Monte Carlo study of core-valence separation schemes

J. Chem. Phys. **108**, 2873–2885 (1998)

The simplicity of variational QMC and the high accuracies obtained when used with explicitly correlated wavefunctions combine to provide a useful and convenient framework for exploring and testing ideas in quantum chemistry. In this paper, the authors report an investigation of the characteristics of core-valence separation schemes with the aim of providing understanding which might be useful especially for heavy elements. The studies were limited to first-row atomic systems, so that errors introduced by core-valence separation could be determined with sufficient accuracies. The bases for comparisons were all-electron VQMC calculations for Li to Ne and their positive ions using single-determinant/Jastrow wavefunctions. The several core-valence partitioning schemes were investigated with modifications of the original function to provide, after identification of two electrons as core electrons, varying degrees of separation as for $\psi = \psi_{core}\psi_{val}$ with and without orthogonality and with and without explicit core-valence correlation. It was shown that orthogonality of core and valence functions is "absolutely necessary for meaningful calculations." With core-valence orthogonality maintained, partitioning could yield good electron distributions and energy differences, provided there was sufficient flexibility within ψ_{core} and ψ_{val}. Under these conditions, the core-electron distribution remained approximately constant and independent of the valence electrons. The best core-valence partitioned wavefunction was found to provide estimates of ionization potentials as accurate as those available from any other method.

112

D. BRESSANINI, M. MELLA, & G. MOROSI

Positronium chemistry by quantum Monte Carlo. I.
Positronium-first row atom complexes

J. Chem. Phys. **108**, 4756–4760 (1998)

Predictions of the energetics of systems containing a positron
(e^+) present a special challenge to most of the conventional methods of quantum chemistry. The prediction of the binding energy of
positronium-atom complexes [A, Ps], where A is an atom and Ps =
$[e^+, e^-]$, is a good example and one for which different theoretical
methods have given widely different results. This paper describes
all-electron (or all-electron and one positron) VQMC and DQMC
calculations to determine the energies of atoms A, their anions A^-
or $[A, e^-]$, and their complexes [A, Ps] or $[A, e^-, e^+]$, where A
= Li, B, C, O, or F. From these energies, the binding energies of
complexes relative to dissociation to [A + Ps] could be predicted.

The calculations were carried out in fixed-node DQMC with
trial functions composed of single determinants for the electrons,
a positron-nucleus term, electron-electron correlation terms, and
electron-positron correlation terms. The electron affinities of the
atoms, given by differences in DQMC-calculated energies for the
atoms and their anions, were found in excellent agreement with
experimental measurements. The binding energies similarly determined for the atom-positronium complexes show the Li, C, O, and
F complexes to be stable and the B complex to be unstable to
dissociation. Other types of electronic structure calculations give a
variety of conflicting energetics. Since the DQMC calculations have
the ability to recover not only electron-electron correlation energy,
but also electron-positron correlation energy, the results "appear
superior in accuracy" to previously published values.

113

M. ZHAO, D. CHEKMAREV, & S. A. RICE

Quantum Monte Carlo simulations of the structure in
the liquid-vapor interface of BiGa binary alloys

J. Chem. Phys. **108**, 5055–5067 (1998)

The liquid-vapor interface of a pure metal has been the subject
of a number of simulations by both molecular dynamics and Monte
Carlo methods. In this work quantum Monte Carlo calculations
for an interface were carried out for electrons and ions in a self-
consistent manner, making use of pseudopotentials for the ions,
a jellium approximation to determine their densities, and a Kohn-
Sham treatment of the electrons. The quantum Monte Carlo aspect
appears in the solution of Kohn-Sham equation, which is the same
as the one-dimensional Schrödinger equation in this application,

$$\left(-\frac{\hbar^2}{2m}\frac{d^2}{dz^2} + V_{\text{eff}}[z, n_e(z)] \right) \psi_n(z) = E_n \psi_n(z),$$

which relates the electron wavefunction ψ_n to the position z normal
to the surface and the effective electron potential V_{eff}, which in turn
depends on the electron density $n_e(z)$. The solution of this equation
for $\psi_n(z)$ and the energy E_n may be determined in variational
QMC by the Metropolis sampling method. With $n_e(z)$ freshly
determined and V_{eff} revised, the process may be repeated, a new
$n_e(z)$ determined, and so forth to achieve self-consistency.

For a pure metal, the development of the ionic interactions and
the pseudopotential is a rather complex procedure, and it is even
more complex for the bismuth-gallium alloy described in this paper.
The end results are structures of the interface expressed in terms
of longitudinal distributions (normal to the surface) of each of the
atomic species, a transverse pair distribution, and the electron densities. Experimental measurements of the structure of alloys such
as these are available from grazing incidence x-ray studies. In this
case, the measured and predicted longitudinal density profiles are
in "rather good" agreement, but differ somewhat in the details.

114

T. YOSHIDA & G. MIYAKO

Quantum Monte Carlo with model potentials for molecules

J. Chem. Phys. **108**, 8059–8061 (1998)

Although simple model potentials lack the solid theoretical basis that justifies nonlocal pseudopotentials, they have been used with moderate success in replacing the core electrons in analytic variational calculations and in diffusion QMC calculations. For this paper, a model potential of long standing, proposed by Bonifacic and Huzinaga[a] in 1974, was investigated in calculations of dissociation energies for the molecules CO, HCl, Na_2, and K_2.

The model potential has an especially simple form for the effective field of the core electrons based on core orbitals. The valence electrons are treated explicitly with Hartree-Fock orbitals used in determinantal trial functions for importance sampling and node location. The numbers of valence electrons treated were: CO, 10; HCl, 8; Na_2, 2; K_2, 2. In each case, the calculated dissociation energy was found to be accurate to within a few percent of the experimental value and much more accurate than either all-electron or model-potential Hartree-Fock calculations. Despite concerns about its justification, the method appears to be surprisingly successful.

[a]V. Bonifacic and S. Huzinaga, Atomic and molecular calculations with the model potential method. I, *J. Chem. Phys.* **60**, 2779 (1974).

115

C.-J. HUANG, C. FILIPPI, & C. J. UMRIGAR

Spin contamination in quantum Monte Carlo
wave functions

J. Chem. Phys. **108**, 8838–8847 (1998)

The Pauli exclusion principle requires that a complete space-spin wavefunction be antisymmetric to the exchange of electrons. This requirement can be met by single-determinant and multi-determinant wavefunctions obtained in SCF and CI calculations, which may then be considered acceptable as trial functions for specifying nodal surfaces and for importance sampling in determining energies. If multiplied by a Jastrow correlation function symmetrical to exchange of electrons, the function remains acceptable. However, in QMC, the Jastrow functions most often used have different coefficients for pairs of electrons of the same and opposite spins. These are not symmetric to all exchanges, and in combination with the determinantal function they give a trial function which is not antisymmetric to exchange of electrons and is not an eigenfunction of the spin operator $\hat{S}^2$. The problem is discussed in detail in this paper. The spin contamination δS^2 of a wavefunction may be readily determined by Metropolis integrations as in variational QMC. The calculations reported for Li and Be show the spin contamination to be very small for unsymmetrical Jastrow terms with the usual coefficients of $\frac{1}{4}$ and $\frac{1}{2}$ for parallel and antiparallel electron pairs. For higher order terms in the Jastrow function, the contamination could be reduced further. With use of symmetric Jastrow factors, the electron-electron cusp conditions are not properly satisfied, so that variance in local energies is higher and computed energies are less accurate, but other properties such as electron densities may be more accurate. An alternative is the use of more complex and computationally expensive functions satisfying both spin and cusp conditions.

116

C. W. GREEFF & W. A. LESTER, JR.

A soft Hartree-Fock pseudopotential for carbon with application to quantum Monte Carlo

J. Chem. Phys. **109**, 1607–1612 (1998)

The popular norm-conserving Hartree-Fock pseudopotentials of Stevens, Basch, and Krauss[a] have been highly successful in reducing the computational effort required in several types of ab initio electronic structure calculations, including both variational and diffusion QMC calculations. These pseudopotentials create a problem specific to diffusion calculations that leads to occasional large fluctuations in local energies and correspondingly higher uncertainties in the calculated energies. In this paper Greeff and Lester propose and describe a new pseudopotential expression for carbon atoms that avoids this problem.

The new pseudopotential is "soft." It is a smooth even function of the electron-nucleus distance r, which is nonzero at $r = 0$ and does not diverge at $r = 0$. It is written in a standard form for use in quantum chemistry programs. Tests with analytic Hartree-Fock calculations showed excellent agreement with HF energies from all-electron calculations and others using standard pseudopotential functions for the small hydrocarbons CH, CH_4, and CH_2CH_2, as well as for the triplet-singlet splitting of CH_2. Tests with the new pseudopotential in VQMC and DQMC calculations for the species C and CH showed a much greater stability, allowed larger time steps without stability problems, and exhibited a much lower time-step dependence than those for earlier pseudopotentials. The C-H bond energy determined from these calculations was in good agreement with the experimental value and with values from other high-level calculations. By making possible the use of larger time steps, these soft pseudopotentials significantly increase the range of practical applications of QMC.

[b]W. J. Stevens, H. Basch, and M. Krauss, Compact effective potentials and efficient shared-exponent basis sets for the first- and second-row atoms, *J. Chem. Phys.* **81**, 6026 (1984).

117

F. SCHAUTZ, H.-J. FLAD, & M. DOLG

Quantum Monte Carlo study of Be$_2$ and group 12
dimers M$_2$ (M = Zn, Cd, Hg)

Theor. Chem. Acc. **99**, 231–240 (1998)

This study was undertaken as a first step toward QMC calculations for clusters of Zn and Cd atoms, following earlier studies of Hg atoms. These dimers are weakly bound in their ground states, but they have some strongly bound excited states which make them interesting candidates for use in eximer lasers, and this has stimulated several successful coupled cluster calculations and spectroscopic studies for these species. As a result, there are sufficient data for comparisons of QMC results.

The QMC calculations were fixed-node diffusion calculations carried out with relativistic energy-consistent large-core pseudopotentials. These ab initio pseuodopotentials eliminated 28, 46, and 78 electrons in Zn, Cd, and Hg, leaving only 2 electrons per atom and 4 for each dimer — like the helium dimer He$_2$. Semi-empirical core-polarization terms were added to the potential energy expression. Relative energies, as well as static dipole polarizabilities, were determined for both atoms and dimers. The polarizabilities required an added finite-difference approach with applied fields. From the complete set of results, the spectroscopic constants R_e, D_e, and ω_e for Zn$_2$ and Cd$_2$ could be derived. As for earlier calculations for Hg$_2$ and for test calculations, these were found to give good agreement with values derived from spectroscopic measurements and from coupled cluster calculations. Similar agreement was found for the polarizabilities.

118

H.-S. LEE, J. M. HERBERT, & A. B. McCOY

Adiabatic diffusion Monte Carlo approaches for studies of ground and excited state properties of van der Waals complexes

J. Chem. Phys. **110**, 5481–5484 (1999)

This paper introduced an "adiabatic" extension to simple diffusion QMC in order to provide (1) a means for generating ψ^2 distributions for evaluating expectation values, and (2) a means for optimizing single-parameter node locations for excited states. The procedures were illustrated with applications to the low-energy vibrational states of the loosely bound van der Waals complexes NeSH and Ar_2HCl.

The adiabatic scheme is based on first-order perturbation theory with a Hamiltonian expressed as $\hat{H} = \hat{H}^{(0)} + \lambda\hat{W}$ and an energy approximated by $E_n = E_n^{(0)} + \lambda E_n^{(1)}$. If a typical diffusion simulation in imaginary time τ is modified by adding to the Hamiltonian a perturbation, $\lambda\hat{W}$, which changes slowly with time as $d\lambda/d\tau$ (the adiabatic approximation), the average energy $E(\tau)$ evolves linearly in λ. This provides the required ψ^2 sampling for expectation values of $\hat{W}$. In the case of a nodal surface which is a function of a single parameter η, the correct node location may be determined by varying the position η with time in a similar fashion. For two simulations, one on each side of a nodal surface, the value of η which produces the same energy for both is the optimized node location. This replaces the use of repeated simulations with η fixed in each.

The calculations were successful in reproducing ground-state expectation values for several interatomic distances and angles determined in analytic variational calculations for each of the species NeSH and Ar_2HCl. For the lowest-energy vibrational states of Ar_2HCl, the nodes in some cases could be defined by symmetry considerations. In others, the node locations were successfully approximated using the adiabatic diffusion method as described, and they produced energies in good agreement with alternate predictions. The advantages of this method arise in the simplification of the search for correct values of the λ and η parameters for expectation values and node location.

119

Y. SHLYAKHTER, S. SOKOLOVA,
A. LÜCHOW, & J. B. ANDERSON

Energetics of carbon clusters C_8 and C_{10} from
all-electron quantum Monte Carlo calculations

J. Chem. Phys. **110**, 10725–10729 (1999)

Carbon clusters in the range C_2 to C_{24} presented a challenge to both experiment and theory long before the discovery of C_{60}, and with that discovery the challenge has been extended to larger systems, particularly to carbon nanotubes. With this paper all-electron fixed-node diffusion QMC calculations were extended to C_8 and C_{10} and compared with those of alternative ab initio methods. For systems of this size, very accurate calculations are available from MP2, MP4, CISD, and CCSD methods. For both C_8 and C_{10}, several isomeric structures lie within 1–2 kcal/mole of each other: linear and cyclic, acetylenic and cumulenic, singlet and triplet. The FN-DQMC calculations produced total electronic energies 0.4 to 1.2 hartrees lower than those of the lowest energy analytic variational and coupled cluster calculations and only 0.2 to 0.4 hartrees above experimentally based total energies. Recovery of correlation energy was 89 to 94%. For C_8 the lowest energy structures were predicted to be the cyclic $C_{4h}\,^1A_g$ structure, the linear $^3\Sigma_g^+$ cumulenic structure, and the linear $^1\Sigma_g^+$ cumulenic structure, all within 10 kcal/mol of each other. For C_{10}, the predicted lowest energy structures were the cyclic $D_{5h}\,^1A_1'$ (distorted cumulenic) and $D_{10h}\,^1A_{1g}$ (cumulenic structures), within 5 kcal/mol of each other. These results are in good agreement with earlier high-level calculations of all types, which in turn are in good agreement with each other. This gives confidence in each of the methods for this size of molecules. The authors included an analysis of the scaling of computation effort with number of electrons N_e, showing a third-power dependence as N_e^3.

120

S. BARONI & S. MORONI

Reptation quantum Monte Carlo:
A method for unbiased ground-state averages and
imaginary-time correlations

Phys. Rev. Lett. **82**, 4745–4748 (1999)

This paper introduces a successful new method for extending diffusion QMC with importance sampling to allow the direct calculation of pure ψ^2 distributions in place of mixed $\psi\psi_T$ distributions. The method removes any bias introduced with a trial function (except that due to node locations) in the calculation of observable quantities which do not commute with the Hamiltonian. It has the advantage, relative to the descendent weighting or forward walking technique,[a] of a controlled walker population which reduces statistical errors.

The authors have named it the reptation quantum Monte Carlo (RQMC) method. Its basic object is a path or sequence of configurations called a "reptile" which may be altered by an action called "reptation," terms adopted from related work in the polymer area. In conventional diffusion QMC with importance sampling walkers are subjected to diffusion, drift, and a local energy–based multiplication to produce a $\psi\psi_T$ distribution. Without multiplication a $\psi_T\psi_T$ distribution is produced. In RQMC multiplication is replaced by a Metropolis rejection procedure, also based on local energies, which produces a $\psi\psi$ distribution of configurations in a reptile. The generation of a new reptile, by removing the tail and adding a new head, is based on diffusion and drift for proposed moves coupled with an acceptance test. Overall, the implementation is quite similar to that for variational QMC.

As described, the method was tested first for the hydrogen atom, followed by the successful demonstration of a calculation for a helium fluid, 64 ^{4}He atoms in a cubic box with periodic boundary conditions. The advantages of RQMC relative to branching diffusion QMC were pointed out for treatment of clusters, films, and superfluids as well as for estimation of electronic forces.

[a]K. S. Liu, M. H. Kalos, and G. V. Chester, Quantum hard spheres in a channel, *Phys. Rev. A* **10**, 303 (1974).

121

M. W. SEVERSON & V. BUCH

Quantum Monte Carlo simulation of intermolecular
excited vibrational states in the cage water hexamer

J. Chem. Phys. **111**, 10866–10874 (1999)

The general problem discussed in this paper is that of predicting vibrational excitation in large hydrogen-bonded clusters, and, as the authors point out, "It should be realized that the problem is very difficult." In this case, the cluster is the cage form of the water hexamer, simplified somewhat by use of rigid-body water molecules. The key to the solution is a new procedure for locating the nodes in the wavefunctions for the excited vibrational states. The method begins with an approximately correct candidate nodal surface, obtained as a zero for a normal coordinate in the harmonic approximation, followed by adjustments to meet the essential criteria for a truly correct nodal surface. These criteria are: (1) energies calculated for either side of a node must be the same, (2) the first derivative of the wavefunction must be continuous across the node, and (3) the wavefunction must be orthogonal to the ground-state wavefunction. It would seem hopeless to make adjustments to satisfy these criteria, but it was discovered they could be met successfully with nodes constructed from normal coordinates, singly in some cases and in linear combinations of up to three coordinates, subject to trial-and-error adjustments of the coefficients. In testing nodal surfaces, the required diffusion QMC calculations were carried out without importance sampling, so that fluxes into the nodal surface from opposite sides could be determined and mismatches eliminated. Despite the complexity of the systems, the energies for ten different low-lying vibrational states were determined. One of these was assigned to the vibration-rotation-tunneling band observed in infrared absorption experiments.[a]

[a] K. Liu, M. G. Brown, and R. J. Saykally, Terahertz laser vibration-rotation tunneling spectroscopy and dipole moment of a cage form of the water hexamer, *J. Phys. Chem. A* **101**, 8995 (1997).

122

S. A. ALEXANDER & R. L. COLDWELL

Relativistic calculations using Monte Carlo methods:
One-electron systems

Phys. Rev. E **60**, 3374–3379 (1999)

The Schrödinger equation is not the only fundamental equation of quantum mechanics that can be solved by Monte Carlo methods. In this paper, a solution of the four-component Dirac equation to determine relativistic energies is reported for the one-electron systems U^{91+}, H_2^+, and Th_2^{179+}. The calculations were carried out in variational QMC, with trial functions optimized by minimizing the variance in local energies.

For a one-electron system with more than one fixed nucleus, the four-component Dirac equation is required. The equation is similar to the Schrödinger equation in that $H\psi = E\psi$, but the wavefunction is a four-component vector with individual components that may be complex, and the Hamiltonian is a 4×4 matrix. The exact solution for a one-electron atom is known, and the authors made use of this knowledge to show that variance minimization would produce correct coefficients in a wavefunction of known functional form. Thus, for U^{91+}, the resulting energy was essentially exact. For the diatomic systems, no exact wavefunctions are known, and three different carefully chosen series functions of up to eight terms were investigated. Among these a mixed wavefunction of Slater atomic orbitals and floating Gaussian orbitals was most successful, although all three gave excellent energies. Nevertheless, for both H_2^+ and Th_2^{179+}, the energies were slightly higher than the best available estimates from other methods of calculation. Properties other than the energy could be evaluated in a similar fashion using the optimized wavefunctions, and additional terms could be added to the Hamiltonian.

123

S. BROUDE, J. O. JUNG, & R. B. GERBER

Combined diffusion quantum Monte Carlo–vibrational
self-consistent field (DQMC-VSCF) method for excited
vibrational states of large polyatomic systems

Chem. Phys. Lett. **299**, 437–442 (1999)

Self-consistent field (SCF) solutions of the Schrödinger equation
can provide approximate, often very accurate, nodes for DQMC
calculations of electronic structure. As shown in this paper, SCF
calculations can do the same in specifying nodes for vibrational
states of polyatomic systems as large as Ar_{13}. The vibrational self-
consistent field (VSCF) method[a] is approximate but quite accurate
in itself, and it has been used previously for prediction of low-level
excitations of protein molecules, water clusters, and other complex
systems. In the calculations described in this paper for Ar_3 and
Ar_{13}, the SCF function was specified as a product of functions
for each of the normal modes, and solutions were obtained with a
sixth-order expansion reducing the required integrals to one dimension. For the argon clusters, the potential energy was specified as
additive pair potentials. The DQMC calculations were performed
using the VSCF nodes as absorbing boundaries without the use
of the wavefunction for importance sampling. In the case of Ar_3,
the DQMC-VSCF results could be compared with nearly exact results from other methods, and they were found better than those of
VSCF alone but with errors as high as 8%. In the case of Ar_{13}, the
excitation energies for DQMC-VSCF and VSCF were found quite
close to each other for the 33 fundamental excited states, thus exhibiting small "correlation" effects. It was noted that the combined
method is better than the VSCF method and that improving on
VSCF as with a vibrational Møller-Plesset theory might provide
even better results.

[a] J. M. Bowman, Self-consistent field energies and wavefunctions for coupled oscillators, *J. Chem. Phys.* **68**, 608 (1978).

124

W. M. C. FOULKES, R. Q. HOOD, & R. J. NEEDS

Symmetry constraints and variational principles in diffusion quantum Monte Carlo calculations of excited-state energies

Phys. Rev. B **60**, 4558–4570 (1999)

This paper addresses the question of the existence of a variational principle for fixed-node QMC calculations. As pointed out, it has often been assumed that energies obtained with fixed nodes are equal to or greater than the energy for an exact wavefunction having the symmetry imposed by the nodes. The usual statement that the energy will be too high if the nodes are wrong is not always correct.

The case of the $2s$ excited state of the hydrogen atom illustrates the problem. A trial node located at a fixed radius r_t greater than the exact node radius r_e divides the space in two regions, one with $r < r_t$ and the other with $r > r_t$. The solution for this inner region has a lower energy than the outer, and a fixed-node calculation started with walkers in both eventually eliminates those in the outer, and produces an energy lower than the true energy for the complete space with a node of $r = r_e$.

As shown in the paper, the statement of the variational principle must be restricted to cases in which the trial function of the fixed nodal surface transforms according to a one-dimensional irreducible representation of the symmetry group of the applicable Hamiltonian. Fortunately, this is often the case, as it is for a determinantal wavefunction which divides the available space into equivalent similar regions. For the ground state of a fermionic system, the variational principle for the full symmetry group does not apply, but for the (fermionic) permutation subgroup the fixed-node variational principle does apply. Thus, a single-determinant trial function meets the requirements for applying the variational principle for many-electron fermionic ground states. It is further shown that the variational principle is not limited to the permutation subgroup, but may also be applied to other subgroups with one-dimensional irreducible representations.

125

X. LIN, H. ZHANG, & A. M. RAPPE

Optimization of quantum Monte Carlo wave functions using analytical energy derivatives

J. Chem. Phys. **112**, 2650–2654 (2000)

The optimization of parameters in trial wavefunctions by minimizing the variance in local energy was used for some of the earliest variational[a] and diffusion[b] QMC calculations, and it has been used ever since by most practitioners of QMC. Perhaps the best known examples are given in the paper by Schmidt and Moskowitz[c] for optimizing their nine-parameter Jastrow functions for first-row atoms. An alternative method of optimization, leading to lower variational energies for these atoms, is reported in this paper. The variational energy itself is minimized using analytical energy derivatives. In this case, the first and second derivatives with respect to coefficients C_m and pairs of coefficients C_m and C_n, $\frac{\partial E}{\partial C_m}$ and $\frac{\partial^2 E}{\partial C_m \partial C_n}$, are determined for a fixed set of configurations sampled using a Metropolis walk. These parameters are then updated to give expected first derivatives as $\frac{\partial E}{\partial C_m} = 0$, and the process is repeated until no further changes occur, usually in three to four iterations. Applied to the atom F, the VQMC energy obtained by Schmidt and Moskowitz of $-99.6685(5)$ hartrees and a correlation energy recovery of 80% was improved to $-99.6792(2)$ hartrees and a correlation energy recovery of 84%. Expanding the Jastrow term to 42 parameters and using the energy optimization procedure gave $-99.6912(2)$ hartrees and 88%. These are significant improvements for variational QMC calculations, and the method is clearly useful for that type of calculation. For diffusion QMC, in which the energy determined depends only on the node structures and the efficiency of the calculation depends on the variance in local energy, the preferred method is likely to be optimization by minimizing the variance.

[a]R. L. Coldwell and R. E. Lowther, Monte Carlo calculation of the Born-Oppenheimer potential between two helium atoms using Hylleraas-type electronic wave functions, *Int. J. Quantum Chem., Symp. Ser.* **12**, 329 (1978).

[b]F. Mentch and J. B. Anderson, Quantum chemistry by random walk: Importance sampling for H_3^+, *J. Chem. Phys.* **74**, 6307 (1981).

[c]K. E. Schmidt and J. W. Moskowitz, Correlated Monte Carlo wavefunctions for the atoms He through Ne, *J. Chem. Phys.* **93**, 4172 (1990).

126

J. B. ANDERSON

Quantum Monte Carlo: Direct calculation of corrections
to trial wave functions and their energies

J. Chem. Phys. **112**, 9699–9702 (2000)

This paper describes a much improved method for reducing the statistical error in fixed-node diffusion QMC calculations. The calculations are designed to determine (1) the difference between the true (fixed-node) wavefunction and a trial wavefunction, and (2) the difference between the fixed-node energy and the expectation value of the energy for the trial wavefunction. When the trial function is a good approximation to the fixed-node result, these differences are small and the statistical errors are smaller still. The nodes of the trial function are not corrected. The general method was described previously.[a] The basic diffusion equation is altered to allow calculation of the difference between the true and trial functions expressed as $g = (\psi - \psi_0)\psi_0$, and this results in additional distributed feed terms for positive and negative walkers. In the new procedure these walkers are generated in Metropolis variational calculations and subsequently used as starting walkers in diffusion calculations with drift. These are continued for a time sufficient to approach equilibrium distributions. In effect, a cancellation of positive and negative walkers in equivalent distributions leaves the correction term g and allows the energy difference to be determined. The method is most useful when the trial function is reasonably accurate. In the case of the ground state of He a trial wavefunction with an energy expectation value of $-2.903\ 724\ 376\ 18$ hartrees led to a correction of $-0.000\ 000\ 000\ 86(2)$ h to give an energy within $0.000\ 000\ 000\ 02$ h of the (separately known) true value. In the case of H_2O with a trial function expectation energy of -75.560 h, the corrected energy value was $-76.16(1)$ h compared to the independently determined fixed-node energy of $-76.17(1)$ h. It remains to be seen whether this can be extended to determining very accurate energy differences in correlated sampling for similar molecules.

[a] J. B. Anderson, Quantum chemistry by random walk: Higher accuracy, *J. Chem. Phys.* **73**, 3897 (1980).

127

A. SARSA, K. E. SCHMIDT, & J. W. MOSKOWITZ

Constraint dynamics for quantum Monte Carlo calculations

J. Chem. Phys. **113**, 44–47 (2000)

In diffusion QMC calculations for clusters of water molecules, it has been found useful in some cases to eliminate the internal motions of the molecules by treating them as rigid rotors.[a] This has greatly simplified the prediction of intermolecular properties of clusters. A similar situation exists for the internal motions of molecules themselves, in which it may be useful to constrain some of the internal motions. Thus, for large molecules one might wish to fix some bond distances or angles which are already tightly, but not completely, restrained. In this paper a convenient method for applying constraints in diffusion QMC is described and demonstrated.

The method is adapted from a procedure[b] used in molecular dynamics simulations in which similar constraints are used. The constraints are applied as distance constraints between specified pairs of atoms by imposing forces to counter the motions which would otherwise alter their separations. Specifying these forces requires a careful consideration of problems of multiplication of time-step errors, but these are successfully eliminated to yield a relatively simple algorithm incorporating importance sampling. At each step, walkers are first moved in the usual way of drift and diffusion without constraint. This is followed by an iterative procedure in which each pair of atoms is adjusted until all constraints are satisfied. The method is demonstrated to give satisfactory results for the rigid-body treatment of a water molecule and for the ground states of methanol with a floppy dihedral-angle rotation involving one of the H–C–O–H strings of atoms. The method is clearly applicable to many different types of molecules, large and small.

[a] V. Buch, Treatment of rigid bodies by diffusion Monte Carlo: Application to the para-H_2-H_2O and ortho-H_2-H_2O clusters, *J. Chem. Phys.* **97**, 726 (1992).

[b] J.-P. Ryckaert, G. Giccotti, and H. J. C. Berendsen, Numerical integration of the cartesian equations of motion of a system with constraints, *J. Comput. Phys.* **23**, 327 (1977).

128

C. FILIPPI & S. FAHY

Optimal orbitals from energy fluctuations in correlated wave functions

J. Chem. Phys. **112**, 3523–3531 (2000)

In conventional Hartree-Fock calculations for molecular systems the optimization of a trial wavefunction by energy minimization with respect to the orbitals of a determinantal wavefunction is relatively easy. When Jastrow factors are included with a determinant as for typical QMC trial functions the task becomes more difficult. Optimization by minimizing the variance in local energies has been one alternative. This paper describes another approach which is, in fact, an analog of the Hartree-Fock method and provides an exact solution to the problem of optimizing the orbitals when a Jastrow function is present.

The method is a hybrid of analytic and Monte Carlo methods. It is iterative in that the orbitals of the Slater function are modified in a series of steps to produce a fully correlated wavefunction which is stationary with respect to variations in the orbitals. Multideterminant wavefunctions can be treated in the same way. The minimization problem is expressed in terms of the Euler-Lagrange equations for a set of one-body operators for the determinant fit to a linear expression for local energies obtained by sampling. This leads to adjustments in the orbitals for the next iteration. Convergence was found to be rapid.

The approach was demonstrated to be successful for the beryllium atom with either single or multiple determinants achieving 98.57% of the correlation energy in variational QMC and 99.76% in fixed-node difffusion QMC. For carbon and neon atoms with pseudopotentials the results were also quite good. Further, for the diamond solid the optimization and a subsequent diffusion calculation produced a value for the cohesive energy of 7.46 ± 0.01 eV/atom which may be compared to an experimental value of 7.37 eV/atom.

129

A. LÜCHOW & R. F. FINK

On the systematic improvement of fixed-node diffusion
quantum Monte Carlo energies using pair natural
orbital CI guide functions

J. Chem. Phys. **113**, 8457–8463 (2000)

With nodes specified by Hartree-Fock wavefunctions, fixed-node DQMC calculations recover typically 93 to 95% of the correlation energy for species like N, N_2, and H_2O. Adding a few more determinants to the wavefunction has been found to give little if any improvement, but it is clear that very large CI functions will give improvement. In this paper the authors report the use of PNOs, pseudo-natural orbitals or pair-natural orbitals,[a] developed in the 1960s for compact multi-determinant wavefunctions in efficient CI calculations with limited computer speeds. With PNOs, transformations are made to produce pairs of orbitals which are either identical or orthogonal, and the CI expansion is much shorter than for the general CI function. With the more compact trial functions, DQMC calculations are much more efficient, and computation requirements are no longer prohibitive. For each of the species N, N_2, and H_2O calculations were carried out with trial functions containing from one to *hundreds* of determinants. Configurations making only small contributions to the calculated energy were deleted. In each case, the recovery of correlation energy is improved to 97 to 98% and this compares well with that of analytic MRCI calculations, but the best nonvariational methods for molecules of this size, such as CCSD(T)-R12, give energies closer to the estimated limits for these species. For large systems, the much more favorable N^3 scaling observed for DQMC may be expected to favor DQMC.

[a]C. Edmiston and M. Krauss, Configuration-interaction calculation of H_3 and H_2, *J. Chem. Phys.* **42**, 1119 (1965).

130

J. C. GROSSMAN, W. A. LESTER, JR., &
S. G. LOUIE

Quantum Monte Carlo density functional theory
characterization of 2-cyclopentenone and
3-cyclopentenone formation from $O(^3P)$ +
cyclopentadiene

J. Am. Chem. Soc. **122**, 705–711 (2000)

One of the myriad reactions of interest in the combustion of
motor fuels is the addition of an oxygen atom in the ^3P state to cyclopentadiene. The generally accepted mechanism for the reaction
begins with addition of the atom to a double bond, and formation
of one or the other of two triplet diradicals depending on the point
of addition. These can both undergo intersystem crossing to form
singlet diradicals which pass through different transition states to
the product, 2- or 3-cyclopentenone. In this study possible reaction
pathways were investigated using a variety of electronic structure
methods to calculate the energetics of critical configurations and
intermediate species, along with the reactants and products. Eight
different theoretical approaches were employed: HF, G2, QMC, and
five variations of density functional theory (LDA, BPW91, BLYP,
B3PW91, B3LYP). The QMC calculations were all-electron variational and fixed-node diffusion calculations with single-reference
trial functions based on natural orbitals. Molecular structures were
optimized for each method, except for diffusion QMC for which
B3LYP structures were used. The QMC results were found to
support the accepted mechanism. Results for other methods were
mixed, with the B3PW91 version of density functional theory in
reasonable agreement with QMC. Both HF and LDA differed from
all others in transition-state barrier heights and heats of reaction.
The G2 method differed for barrier heights but agreed with QMC
for heats of reaction. The authors provided an extended discussion of the advantages and disadvantages of the several methods
and concluded that a combination of B3LYP or B3PW91 calculations for geometries with single-point diffusion QMC calculations
for energies would provide accuracies of ~1–2 kcal/mol in potential
energy surfaces for similar reactions.

131

F. PEDERIVA, C. J. UMRIGAR, & E. LIPPARINI

Diffusion Monte Carlo study of circular quantum dots

Phys. Rev. B **62**, 8120–8125 (2000)

Several QMC calculations had earlier treated quantum dots of various shapes and sizes by either variational or diffusion methods. This study extended much farther, with ground and low-lying excited states for circular two-dimensional dots with 2 to 13 electrons. The method was fixed-node diffusion QMC with importance sampling. Comparisons were made with Hartree-Fock and density functional calculations at the LSDA level.

The system treated was a standard circular quantum dot with N electrons confined in two dimensions by a parabolic potential $V = \frac{1}{2}kr^2$ centered at the origin. The masses, dielectric constants, and the potential energy constants were specified as those for a GaAs crystal. Atomic units were replaced by effective atomic units. Anti-symmetry rules were the same as for atomic systems. Trial functions were expressed for each case as sums of one to five Slater determinants from density functional calculations, along with elaborate generalized Jastrow functions. The coefficients required for satisfying cusp conditions were different from those usually found for three-dimensional problems. A total of 23 cases with different numbers of electrons and L, S eigenstates were treated. The first version of the paper contained a few errors. Corrected results (also with lower uncertainties) were reported in a second version.[a] In all cases the diffusion QMC energies were lower than the variational QMC energies, which were in turn lower than the HF energies. Fixed-node error was expected to be small. The differences from LSDA density functional energies were small but not negligible.

[a] F. Pederiva, C. J. Umrigar, and E. Lipparini, Diffusion Monte Carlo study of circular quantum dots, *Phys. Rev. B* **68**, 089901 (2003).

132

D. C. CLARY

Torsional diffusion Monte Carlo: A method for quantum simulations of proteins

J. Chem. Phys. **114**, 9725–9732 (2001)

Before this paper studies of the molecular dynamics of proteins had been carried out primarily with classical dynamics. This paper shows that quantum-dynamical effects are important in determining the energies and geometries of typical proteins. The examples in this study are the proteins gelsolin (with 356 atoms and 142 torsions) and gp41-HIV (1101 atoms and 452 torsions), which were treated with all-atom force fields in diffusion QMC with restriction to torsional motions only. These motions have a much lower frequency than other motions for proteins, and they are the essential motions determining the dynamics of protein folding. The authors were able to assemble a general code for torsional diffusion QMC calculations to be carried out automatically for any protein with known connectivity or atom coordinates converted to internal coordinates of fixed-bond angles, fixed bond lengths, and variable torsion angles. All-atom potential energies were obtained from standard force-field programs. The torsional diffusion calculation is essentially identical to one-dimensional translational motion calculations, except that angles replace distances, and the $\hbar^2/2m$ is replaced by $\hbar^2/2I$ in the diffusion terms. The results obtained from the calculations are the zero point energies and, with use of descendent weighting methods, distributions of structures or simply vibrationally averaged structures. The calculations show structures significantly changed from minimum-energy structures. It is noted that binding energies for drugs that might inhibit functions of such proteins are likely to be affected and that QMC calculations of this type may be important in drug discovery.

133

S. MANTEN & A. LÜCHOW

On the accuracy of the fixed-node diffusion quantum
Monte Carlo method

J. Chem. Phys. **115**, 5362–5366 (2001)

In this work Manten and Lüchow assessed the accuracy of all-electron fixed-node diffusion QMC calculations with nodes taken from Hartee-Fock wavefunctions with high-quality basis sets. The electronic energies associated with 17 different reactions involving a total of 20 small molecules were determined and compared with experimental values and with those calculated by Klopper et al.[a] using the coupled cluster method CCSD(T) with correlation consistent basis sets cc-pVDZ and cc-pVTZ. The molecules involved were limited to those containing the atoms H, C, O, N, and F. Typical reactions were $H_2CO + 2H_2 \rightarrow CH_4 + H_2O$ and $H_2O + F \rightarrow HOF + H$. The geometries for the QMC calculations were optimized with MP2/cc-pVTZ calculations. The QMC total energies for the 20 molecules were lower than CCSD(T) values, but only valence electrons were correlated in the CCSD(T) calculations. Node location error was approximately 0.0015 hartrees for the C, N, O, and F atoms and their hydrides, but substantial cancellation of node error is to be expected in determining enthalpy changes for reactions.

The results indicate the FN-DQMC method is more accurate than the CCSD(T)/cc-pVDZ method and nearly as accurate as the CCSD(T)/cc-pVTZ method. The differences from experimental reaction enthalpies were typically 2 to 5 kcal/mol for the FN-DQMC calculations. The exception was a reaction involving ozone, for which an MCSCF wavefunction was necessary for suitable specification of the nodes.

[a]W. Klopper, K. L. Bak, P. Jorgensen, J. Olsen, and T. Helgaker, Highly accurate calculations of molecular electronic structure, *J. Phys. B* **32**, R103 (1999).

134

A. J. WILLIAMSON, R. Q. HOOD, & J. C. GROSSMAN

Linear-scaling quantum Monte Carlo calculations

Phys. Rev. Lett. **87**, 246406-1/4 (2001)

This study demonstrates linear scaling with system size in fixed-node diffusion QMC calculations with pseudopotentials for carbon fullerenes and for hydrogenated silicon clusters with nearly 1000 valence electrons. The range covered is C_{20} to C_{180} and SiH_4 to $Si_{211}H_{140}$. The calculations were carried out with standard pseudopotentials (leaving four electrons per C or Si atom) and Slater-Jastrow trial functions with orbitals from LDA density functional calculations. The key elements leading to linear scaling were constructions of sparse determinants and simplification of orbital functions. The LDA orbitals were obtained with a plane-wave basis set which was transformed to yield Wannier functions with maximum localization. These functions were then fitted to cubic spline expressions with a limited number of grid points and cut-off distances matching those of the plane-wave orbitals. The computational effort for evaluating trial functions of this type was clearly seen to behave in a linear fashion. Exploration of the effects of variation of cut-off radius revealed that energies, sensitive only to node structures, increased rapidly for cut-offs less than 7 bohr but were unaffected for cutoffs greater than 7 bohr. The accuracies of the calculations were checked in other ways. These included comparison calculations for SiH_3 and Si_5H_{12} using Gaussian, original plane-wave, and Wannier function basis sets which gave close agreement for energies.

For the largest molecule, $Si_{211}H_{140}$, an analysis of the algorithm showed 90% of the effort scaled as N, and 10% scaled as N^2 or N^3. It was noted that most of the latter could be shifted to N scaling with minor alterations. Thus, linear behavior to several thousand electrons could be expected.

135

S. B. HEALY, C. FILIPPI, P. KRATZER, E. PENEV, & M. SCHEFFLER

Role of electronic correlation in the Si(100) reconstruction: A quantum Monte Carlo study

Phys. Rev. Lett. **87**, 016105-1/4 (2001)

The 100 surface of a silicon crystal is believed to reconstruct to form rows of silicon dimers in either a "symmetric" pattern (| | | | | | | |) or a somewhat deformed "buckled" pattern (/\/\/\/\) in which the dimers are twisted alternately within each row. At room temperature the dimers may flip back and forth between configurations and they appear symmetric in scanning tunneling microscopy. Below 120 K the buckled configuration appears, but below 20 K the symmetric configuration has been indicated but might be due to flipping, and arguments in favor of each of the configurations have been made. Theoretical work is equivocal. Density functional (DFT) calculations favor the buckled configuration. Multiconfiguration self-consistent field (MCSCF) and configuration interaction (CI) calculations favor the symmetric. The indicated differences in energies for the two structures are small, and electron correlation (note the title) is thus expected to be important in determining the minimum-energy structure.

Diffusion QMC provides the required accurate description of correlation for large systems. As in the earlier studies with other methods the authors treated the model systems Si_9H_{12}, $Si_{15}H_{16}$, and $Si_{21}H_{20}$ — clusters which have one, two, and three dimers in configurations like those on the Si(100) surface. The calculations were made using silicon pseudopotentials tested in calculations for Si_2. For the intermediate cluster the energies of the symmetric and the buckled configurations were nearly identical. For the larger cluster, with three dimers, the buckled configuration was favored by 0.34(6) eV or 0.11(2) eV per dimer. This energy difference is similar to that for DFT calculations (LDA to B3LYP). The agreement suggests that the dynamic correlation effects, described better in DQMC and DFT than in MCSCF and CI, are important. The buckled configuration is thus to be favored, but many questions in both theory and experiment remain unresolved for this system.

136

R. BAER

Ab initio computation of molecular singlet-triplet energy differences using auxiliary field Monte Carlo

Chem. Phys. Lett. **343**, 535–542 (2001)

About 15 years before this paper was published, all-electron fixed-node diffusion QMC calculations by Reynolds, Dupuis, and Lester[a] had successfully duplicated the 9 kcal/mol seen in direct experimental measurements of the classic problem of singlet-triplet splitting for the methylene molecule CH_2. This second QMC paper on the subject was designed less to investigate the CH_2 system and more to investigate the method, in this case the shifted-contour auxiliary field Monte Carlo (SC-AFMC) method,[b] which overcomes some of the stability problems of the original AFMC method.[c] The paper describes the application of SC-AFMC to the simplest singlet-triplet problem of linear H-He-H as well as that of methylene, this time with pseudopotentials.

The applications were far from straightforward and required investigation of the various approximations needed to obtain the energy differences. Nonlocal B-LYP pseudopotentials were used, along with combinations of restricted and unrestricted Hartree-Fock determinantal wavefunctions. Correlated sampling was successful in reducing statistical error in energy differences but convergence was slow. Near degeneracies led to problems for H-He-H. Results were sensitive to cell size and grid spacing and especially sensitive to the type of pseudopotential used.

Nevertheless, suitable choices produced singlet-triplet splittings in agreement with accepted values when pseudopotentials of the generalized gradient type were used. However, this agreement might have been fortuitous. Further studies of alternative approaches were indicated.

[a] P. J. Reynolds, M. Dupuis, and W. A. Lester, Jr., Quantum Monte Carlo calculation of the singlet-triplet splitting in methylene, *J. Chem. Phys.* **82**, 1983 (1985).

[b] R. Baer, M. Head-Gordon, and D. Neuhauser, Shifted-contour auxiliary field Monte Carlo for *ab initio* electronic structure: Straddling the sign problem, *J. Chem. Phys.* **109**, 6219 (1998).

[c] G. Sugiyama and S. E. Koonin, Auxiliary field Monte Carlo for quantum many-body ground states, *Ann. Phys. (N.Y.)* **168**, 1 (1986).

137

J. C. GROSSMAN

Benchmark quantum Monte Carlo calculations

J. Chem. Phys. **117**, 1434–1440 (2002)

In this paper Grossman reported pseudopotential fixed-node diffusion QMC calculations of atomization energies for the G1 set of 55 small molecules,[a] a standard benchmark set for testing methods in quantum chemistry. To maintain the uniformity in the calculations, they were carried out with the same basis sets and orbital types for single-determinant trial functions and effective core potentials (with a few exceptions). Variational QMC was used to optimize the determinant-Jastrow trial functions. The mean absolute deviation E_{MAD} in atomization energy from the experimental values was found to be 2.9 kcal/mol, the maximum deviation was 14 kcal/mol (for SO_2), and the next five largest deviations were 7–8 kcal/mol.

A comparison with results from other methods tested with the G1 set shows the FN-DQMC method to be competitive. For E_{MAD} the list is: FN-DQMC, 2.9 kcal/mol; DFT (LDA), 40 kcal/mol; DFT (B3LYP or B3PW91), 2.9 kcal/mol; CCSD(T)/aug-cc-pVQZ, 2.8 kcal/mol; CCSD(T)/extrapolated, 1.3 kcal/mol; semi-empirical G1 theory, 1.6 kcal/mol.

Additional calculations with different trial functions indicated that node location errors could be significantly reduced with multideterminant trial functions. Computational effort was not compared, but linear scaling for FN-DQMC was mentioned as an important advantage for QMC.

[a] J. A. Pople, M. Head-Gordon, D. J. Fox, K. Raghavachari, and L. A. Curtiss, Gaussian-1 theory — A general procedure for prediction of molecular energies, *J. Chem. Phys.* **90**, 5622 (1989).

138

F. SCHAUTZ & H.-J. FLAD

Selective correlation scheme
within diffusion quantum Monte Carlo

J. Chem. Phys. **116**, 7389–7399 (2002)

This paper describes the development of a new scheme for using an effective model potential to reduce the number of electrons explicitly treated in a QMC calculation for a molecular system. Like other pseudopotential and model-potential approaches it is based on differing treatments of core and valence (or frozen and explicitly correlated) electrons. It differs from the earlier schemes of Yoshida and co-workers[a] in that it is adapted to multicenter systems and includes a projectionlike operator to enforce orthogonality between orbitals for the two sets of electrons. The model potential includes Coulomb and exchange terms and the projectionlike operator. The latter two are nonlocal. For these a localization procedure, similar to those used in pseudopotential calculations, was derived. In implementing the scheme the authors had to explore a number of questions, such as those related to nonequivalent nodal domains, and to devise a number of approximations to simplify the calculations.

The triple-bonded nitrogen molecule N_2, with correlation for the electrons of the bonding orgitals and with the nonbonding orbitals frozen, was used as a test case. A coupled cluster [CCSD(T)] calculation with a large basis set provided a benchmark with a correlation energy of 228 mhartree for the triple bond estimated to be accurate within a few mhartree. The corresponding result for the QMC calculation using the best approximations differed by 2–6 mhartree. Thus, quantitative agreement was obtained within the limits of statistical error. The authors indicated a "good chance" for further development to enable efficient treatment of problems involving solids and surfaces.

[a]T. Toshida, Y. Mizushima, and K. Iguchi, Electron affinity of Cl: A model potential-quantum Monte Carlo study, *J. Chem. Phys.* **89**, 5815 (1988).

139, 140

D. BLUME

Fermionization of a bosonic gas under highly elongated confinement: A diffusion quantum Monte Carlo study

Phys. Rev. A **66**, 053613-1/7 (2002)

G. E. ASTRAKHARCHIK & S. GIORGINI

Quantum Monte Carlo study of the three- to one-dimensional crossover for a trapped Bose gas

Phys. Rev. A **66**, 053614-1/6 (2002)

These two papers, published back to back, provide superb examples of the use of a quantum Monte Carlo method to obtain exact solutions for a problem difficult to treat by other methods. The problem is that of a bosonic atomic gas confined in a cigar-shaped trap. As the cigar is narrowed the atoms are less able to pass each other, their behavior approaches that of fermions, and they are said to exhibit "fermionization." Using diffusion QMC the authors were able to determine the ground-state energies, examine structural properties, confirm that fermionization can occur, and predict conditions under which it might be observed.

The calculations by Blume were carried out for 2–20 atoms confined in axially symmetric traps of varied aspect ratios. The confining potential for each atom was given by $V_k = c_\rho \rho_k^2 + c_z z_k^2$ (radial and axial positions squared). The atom-atom interactions were simple hard-sphere potentials. The (nodeless) guided diffusion calculations used a trial wavefunction of the Jastrow type, a product of atom-trap terms and atom-atom terms.

The calculations by Astrakharchik and Giorgini were carried out similarly but included both hard and soft spheres with numbers in the range of 5 to 100 atoms. Comparisons were made with predictions of the Lieb-Liniger equation of state.

The results of each study clearly show fermionization behavior and the approach to the 1-D Tonks-Girardeau limit. For a small aspect ratio (wide) the behavior is that of bosons. For a large aspect ratio (narrow) the behavior is that of fermions. The observed trends suggest conditions most promising for experimental observations.

141

K. E. RILEY & J. B. ANDERSON

Higher accuracy quantum Monte Carlo calculations of the barrier for the H + H$_2$ reaction

J. Chem. Phys. **118**, 3437–3438 (2003)

For many QMC calculations, the accuracy of the results may be improved by simply extending the calculation time to increase the number of samples. With an increase in computer speed of about a factor of ten every four years, and an inverse square-root dependence of statistical error on number of samples, one should expect a factor-of-ten improvement in accuracy every eight years. That is the case for the calculations reported in this paper. An "exact" QMC calculation[a] of the potential energy surface for the reaction H + H$_2$ in 1992 gave a barrier height of 9.61 ± 0.01 kcal/mol. The same program was run with only minor changes on a faster machine about ten years later to give the results in this paper. This time, the barrier height was found to be 9.608 ± 0.001 kcal/mol. Energies for other points on the potential energy surface were similarly improved in accuracy. These included the energies of points for the ground state and the first excited state near the Jahn-Teller cusp.

[a]D. L. Diedrich and J. B. Anderson, An accurate Monte Carlo calculation of the barrier height for the reaction H + H$_2$ → H$_2$ + H, *Science* **258**, 786 (1992).

142

S. MANTEN & A. LÜCHOW

Linear scaling for the local energy in quantum Monte Carlo

J. Chem. Phys. **191**, 1307–1312 (2003)

This paper describes the use of localized orbitals and short-range correlation functions to achieve near linear scaling in fixed-node diffusion QMC calculations. The success of the method is demonstrated with a series of all-electron calculations for linear hydrocarbons in the range of C_2H_6 to $C_{25}H_{52}$ with 18 to 202 electrons.

The procedure used was not very different from that of earlier all-electron calculations for systems of these sizes. Some changes were needed for evaluation of the trial wavefunctions and their derivatives. The molecular orbitals for the determinantal wavefunction were composed of contracted Gaussian functions decaying rapidly with distance but limited by a cut-off radius. The resulting determinants were sparse and could be evaluated by a standard matrix LU decomposition package. The standard Schmidt-Moskowitz correlation function was modified with use of a short-range scaled distance to allow short cut-off radii.

In these fixed-node calculations the energy was dependent only on the location of the nodes as specified by the trial function. The minimum cut-off distances required to avoid cut-off effects were determined in variational QMC calculations, a much stricter requirement. In general, shorter cut-off lengths increased the variance in the local energy so that a balance of variance vs. computation speed was required. Consideration of all effects led to a prediction of linear scaling shifting to quadratic for very large systems, but linear scaling was found in effect for the calculations all the way up to $C_{25}H_{52}$.

143

J. CARLSON, J. MORALES, JR.,
V. R. PANDHARIPANDE, &
D. E. RAVENHALL

Quantum Monte Carlo calculations of neutron matter

Phys. Rev. C **68**, 025802-1/13 (2003)

Pure neutron matter is not likely to be found in any of the Earth's laboratories, but it has been found to exist as the material of neutron stars. It presents an interesting and difficult many-body problem which may be attacked by QMC methods. In this paper the authors report variational, approximate fixed-node, and released-node QMC calculations for uniform neutron matter in a three-dimensional box with periodic boundary conditions. These allow the determination of the energy $E(\rho)$ as a function of the neutron density ρ for a realistic two-neutron interaction expression. The results are used to assess the accuracy of earlier variational approaches.

The system treated was that of 14 neutrons in a cube at densities in the range of 1/4 to 3/2 the density ρ_0 typical for nuclear matter. Most of the calculations were done for the Argonne $v8'$ two-body interaction potential. The trial wavefunction was a correlated Slater function, identical for the variational and diffusion calculations, specifying the nodes and used in importance sampling. Corrections were made for the effects of finite box dimensions. The released-node calculations were limited to very short time intervals especially for the higher densities.

The energies $E(\rho)$ found in the diffusion QMC calculations were 5 to 10% lower than those of the variational QMC. The released-node results, with relatively large statistical errors, were not significantly different from the fixed-node results. The energies determined for the lower densities were estimated to be accurate within 2% and those for the higher densities somewhat less accurate. The results showed earlier predictions to be generally correct. Comparisons of results from variational chain summations (VCS) indicated an overall accuracy of about 10% for that method.

144

C. A. SCHUETZ, M. FRENKLACH, A. C. KOLLIAS, & W. A. LESTER, JR.

Geometry optimization in quantum Monte Carlo with solution mapping: Application to formaldehyde

J. Chem. Phys. **119**, 9386–9392 (2003)

Since analytic gradients of the energy with respect to nuclear positions are not easily obtained in QMC calculations, determining the locations of critical points, such as energy minima and saddle points, on potential energy surfaces is not straightforward. The statistical uncertainty in QMC energies prevents the direct estimation of these derivatives, and although several techniques for overcoming the problem have been developed their applications have been limited to very small systems. This paper describes a different approach, more readily applicable, in which the number and location of points in the vicinity of a critical point are chosen in such a way as to minimize the computation effort required in predicting the location of the critical point. The method is generally known as "solution mapping" and is often referred to as "factorial design." The procedure is illustrated with calculations to determine the minimum energy or equilibrium geometry of formaldehyde H_2CO, assumed planar and symmetric about the C-O bond, in terms of the C-O and C-H bond lengths and the H-C-O angle. The potential energies calculated at specified points are fit to a polynomial expression of ten terms involving the three variables as $1, x_1, x_2, x_3, x_1^2, x_2^2, x_3^2, x_1x_2, x_1x_3, x_2x_3$, and ten coeffcients. With 15 points prescribed by the factorial design the determination of the coefficients is simplified by the column orthogonality of the design matrix. Of course, statistical uncertainties in the energies lead to uncertainties in the coefficients, but the method minimizes their effects.

For the case of formaldehyde, optimizations were made for both variational and diffusion QMC for two different high-quality trial wavefunctions with effective core potentials. The several calculations were entirely successful in locating the equilibrium structures. For the DQMC calculations the two bond lengths, the bond angle, and their vibrational frequencies agreed within their uncertainties with experimentally determined values.

145

M. NEKOVEE, W. M. C. FOULKES, & R. J. NEEDS

Quantum Monte Carlo studies of density functional theory

Math. Comput. Simulat. **62**, 463–470 (2003)

It is difficult to imagine that density functional theory (DFT) could be useful in improving quantum Monte Carlo methods, but it is easy to imagine that QMC could be useful in improving density functional methods. Since QMC can produce accurate wavefunctions and since wavefunctions lead to all other properties one should be able, *in principle,* to use QMC to predict the exchange-correlation energy functional $E_{xc}[n(r)]$, the key (unknown) quantity required in DFT. *In practice,* that is not so easy, but this paper shows how QMC can take steps in that direction by providing insights gained from comparisons of results from QMC and DFT.

The authors considered the case of the strongly inhomogeneous electron gas using variational QMC to obtain wavefunctions ψ^λ and energies $E[\psi^\lambda]$ for the ground state of the Hamiltonian $\hat{H}^\lambda = \hat{T} + \lambda \hat{V}_{ee} + \hat{V}^\lambda$ associated with the coupling constant λ in the range 0 to 1. Optimization of a Slater-Jastrow many-body wavefunction was made by minimizing a penalty function given by a weighted sum of the variance in local energies and a sum of squares of electron density differences. The variational parameters in ψ^λ and V^λ were repeatedly varied to reach convergence followed by additional calculations to obtain the local values of the exchange-correlation energies e_{xc}^λ and the exchange-correlation holes n_{xc}^λ from which E_{xc} and n_{xc} could be determined.

Analysis of differences in e_{xc} for the local density approximation LDA, the generalized gradient approximation GGA, and QMC provided insights into the differing behavior of DFT for the electron gas, for solids, and for atomic systems. The results also suggested possible directions for improving the exchange-correlation functional.

146

W. SCHATTKE, R. BAHNSEN, & R. REDMER

Variational quantum Monte-Carlo method
in surface physics

Prog. Surf. Sci. **72**, 87–116 (2003)

This paper is of interest for its discussions of the QMC treatment of lattice displacements in bulk matter and at solid surfaces with the aim of the prediction of phonon spectra. Earlier calculations had been made with density functional methods. The authors report variational QMC calculations of ground-state energies for gallium arsenide crystals using both local and nonlocal pseudopotentials coupled with trial wavefunctions of the Slater-Jastrow type. Shifting of atomic positions and their corresponding orbitals allowed the simulation of phonon displacements as well as the determination of the bulk modulus and related properties.

As an example, the case of the [111] phonon for bulk GaAs was treated in a (2,2,8) simulation cell with arsenic atoms fixed and gallium atoms displaced by up to 10% of the lattice constant. The energies obtained were fit to a quadratic function of the displacement from which the frequency of the mode considered could be calculated. Agreement with experimental values for the frequency was found within the estimated error of a few percent.

For the surface the general scheme was the same, but a number of changes were required for a slab geometry. A "confinement potential" was added, the treatment of periodicity was altered, and the Ewald summations were altered. The calculated energies and electron densities revealed several inconsistencies, such that the surface calculations could "only be considered as a first step." Nevertheless, the calculations demonstrate that QMC treatment of surfaces, including phonon dynamics at surfaces, is feasible.

147

A. ASPURU-GUZIK, O. EL AKRAMINE, J. C. GROSSMAN, & W. A. LESTER, JR.

Quantum Monte Carlo for electronic excitations of
free-base porphyrin

J. Chem. Phys. **120**, 3049–3050 (2004)

With 216 electrons, the free-base porphyrin $C_{20}N_4H_{14}$ presents a challenge to ab initio electronic structure calculations. Since it is at the center of the mechanism of photosynthesis, as well as several other biological processes, there has been much interest in this molecule and its electronic spectrum, but despite a number of earlier studies, important questions remain open. This paper provides the most accurate calculations to date of energy levels of states involved in several allowed and nonallowed transitions seen in the spectrum. These were obtained in all-electron fixed-node diffusion QMC calculations for the singlet ground states, the first excited singlet state $1\ ^1B_{2u}$, and the lowest triplet state $1\ ^3B_{2u}$. The trial functions for importance sampling were based on Hartree-Fock and modified Hartree-Fock determinants using a 6-311G** basis set and Jastrow correlation factors. The calculations were carried out for thousands of hours to reduce statistical error to about 0.1 eV. At $-988.985(3)$ hartrees, the computed ground state energy was 5.6 hartrees lower than the Hartree-Fock energy and 2.3 hartrees lower than the coupled cluster CCSD(T) energy. The computed excitation energies were in good agreement with experimental values: for $^1B_{2u}$, 2.45(8) eV versus experimental values of 2.42 and 2.46 eV; for $^3B_{2u}$, 1.60(10) eV versus an experimental value of 1.58 eV. These clearly demonstrate the suitability of diffusion QMC for predicting energetics of large biological molecules.

148

P. CAZZATO, S. PAOLINI, S. MORONI, & S. BARONI

Rotational dynamics of CO solvated in small He clusters: A quantum Monte Carlo study

J. Chem. Phys. **120**, 9071–9076 (2004)

The reptation QMC method introduced by Baroni and Moroni[a] a few years earlier is demonstrated in this paper to be the method of choice for predicting the rotational-vibrational features of the infrared spectrum of CO in small helium droplets. The calculations gave good agreement with experimental measurements for these relatively complex systems and provided insights into some of their characteristics not observable in experiments. The calculations were carried out for clusters containing one CO molecule and up to 30 He atoms. The CO molecule was treated as rigid, and He-He and He-CO interactions were treated as pairwise additive, with pair terms from ab initio calculations. The trial wavefunction for this nodeless bosonic system was a Jastrow function made up of exponential functions dependent on the CO distance and orientation with respect to He atom and He-He distances.

The reptation QMC calculations were carried out as diffusion QMC calculations with importance-sampling drift, but without multiplication so that the energy obtained was the expectation value of the energy for the trial wavefunction. The reptation part of the calculation produced exact (or nearly exact) expectation values and, more important in these calculations, unbiased quantum correlation functions from which the absorption spectra of the CO molecule could be determined. The results exhibit a variety of interesting phenomena. The binding energy for an additional atom first increases and then decreases before leveling out with increasing cluster size. The calculated rotational absorption spectra show two types of lines, one increasing in strength with cluster size, the other decreasing with cluster size. Much more detail, especially in regard to the droplet shape, revealed in the calculations.

[a]S. Baroni and S. Moroni, Reptation quantum Monte Carlo: A method for unbiased ground-state averages and imaginary-time correlations, *Phys. Rev. Lett.* **82**, 4745 (1999).

149

F. SCHAUTZ, F. BUDA, & C. FILIPPI

Excitations in photoactive molecules from quantum Monte Carlo

J. Chem. Phys. **121**, 5836–5844 (2004)

Among the several quantum approaches to the accurate description of excitation processes in biological systems, the most promising candidates are: CASPT2, complete active space second-order perturbation theory; TDDFT, time-dependent density functional theory; ROKS, the restricted open-shell Kohn-Sham method; and QMC, quantum Monte Carlo. Each of these has its disadvantages. CASPT2 scales poorly with system size and is limited to smaller molecules. ROKS and TDDFT can be applied to very large systems but they may or may not be adequate for photoactive molecules. QMC scales favorably with system size but is computationally expensive. The authors address these issues in this paper with fixed-node diffusion QMC calculations for ground and excited states for protypical photosensitive molecules, and companion calculations by CASPT2, TDDFT, and ROKS methods.

The species chosen were formaldimine (CH_2NH), formaldehyde (CH_2O), and a protonated Schiff base ($C_5H_6NH_2^+$) serving as a model for the retinal chromophore. The all-electron calculations were carried out with trial functions made up of linear combinations of Slater determinants, with Jastrow correlation factors optimized to minimize the average of energies for ground and excited states, using a fixed set of orbitals available for both. This was found to be important in obtaining accurate excitation energies. Transitions were examined along isomerization paths for each of the species, and comparisons of results were made for a variety of configurations.

Important differences in several of the predictions were observed. QMC and TDDFT results were in general agreement except for differences in the isomerization path for the Schiff base model. ROKS was found to exhibit large differences for lower symmetry structures. A limited number of comparisons showed the CASPT2 and QMC results to be qualitatively similar but with lower excitation energies for CASPT2.

150

K. HONGO, R. MAEZONO, Y. KAWAZOE, H. YASUHARA, M. D. TOWLER, & R. J. NEEDS

Interpretation of Hund's multiplicity rule for the carbon atom

J. Chem. Phys. **121**, 7144–7147 (2004)

Originating from experimental observations and published in 1925, Hund's rule[a] states that the electronic state with the lowest energy is the one with the highest spin multiplicity S. Obeyed by most atoms and molecules, although exceptions are known, the rule has been explained (incorrectly) as due to the lower average electron-electron repulsion V_{ee} for electrons of the same spin in higher S states, and (correctly)[b] as due to lowering of the electron-nucleus energy V_{en} as electrons of the same spin avoid each other and can move closer to the nucleus. The authors call this Boyd's *less screening*[c] mechanism. A completely satisfactory analysis of the phenomenon requires very accurate wavefunctions, and in this paper the authors used VQMC and DQMC calculations to provide those wavefunctions for singlet and triplet states of the carbon atom.

The trial wavefunctions used were single-determinant functions with 12-parameter Jastrow terms which recovered 88% of the correlation energy in DQMC. For both singlet and triplet VQMC and DQMC, the virial ratio $-V/T$ was within statistical error (0.001 or 0.002) of the exact value of 2. The triplet energy obtained was 0.0644(7) hartrees lower. For analysis of average values V, T, V_{ee}, and V_{en}, the short linear extrapolation based on ψ_t^2 to $\psi_t\psi$ to ψ^2 was used. These gave $V_{ee}^{S=1} > V_{ee}^{S=0}$, $V_{en}^{S=1} < V_{en}^{S=0}$, and $T^{S=1} > T^{S=0}$ and confirmed the "less screening" explanation of Hund's rule. It was suggested that extending the study to heavier atoms would not be difficult.

[a]F. Hund, Zur Deutung verwickelter Spektren, *Z. Phys.* **33**, 345 (1925); **34**, 296 (1926).

[b]E. R. Davidson, Single-configuration calculations on excited states of helium, *J. Chem. Phys.* **41**, 656 (1964); **42**, 4199 (1965).

[c]R. J. Boyd and C. A. Coulson, Coulomb hole in some excited states of helium, *J. Phys. B* **6**, 782 (1973); **7**, 1805 (1974).

151

D. L. CRITTENDEN, K. C. THOMPSON, M. CHEBIB, & M. J. T. JORDAN

Efficiency considerations in the construction of interpolated potential energy surfaces for the calculation of quantum observables by diffusion Monte Carlo

J. Chem. Phys. **121**, 9844–9854 (2004)

This paper addresses the problem of choosing configurations of a loosely bound molecular complex for constructing a potential energy surface to be used in calculating quantum observables of the complex. For a tightly bound system, diffusion QMC has been shown to be an efficient sampling method for this purpose.[a] A loosely bound complex presents somewhat different problems, and the authors investigated alternative procedures which they found useful in combination with diffusion QMC sampling. The example chosen is the very floppy prototypical water dimer, for which a suitable potential energy surface was available, and this surface took the place of ab initio calculations in providing potential energies for configurations sampled. The data obtained in sampling were not fit to a function representing the surface, but were used in interpolation to obtain the values of potential energies at other points on the surface required in calculations of quantum observables.

The interpolations were accomplished with a Taylor series expansion, with higher weights for nearby data points. An initial potential energy surface, or set of data points, was chosen by diffusion QMC sampling. This set was improved by adding data points from additional QMC calculations, as well as from classical trajectory calculations with selections biased toward those with energies differing most from predictions using the original data set. The effects of varying sizes and types of sampling were investigated in detail. A two-part scheme with a combination of sampling types was found most efficient. With this scheme, approximately 50 points were required for converged zero-point energies and about 520 points for converged distributions of distances separating the two oxygen atoms.

[a]R. P. A. Bettens, Bound state potential energy surface construction: Ab initio zero-point energies and vibrationally averaged rotational constants, *J. Am. Chem. Soc.* **125**, 584 (2003).

152

N. GOLDMAN & R. J. SAYKALLY

Elucidating the role of many-body forces in liquid water. I. Simulations of water clusters on the VRT(ASP-W) potential surfaces

J. Chem. Phys. **120**, 4777–4789 (2004)

In the quest for a quantitative simulation of liquid water it appears the potential energy of the interaction of water molecules converges very rapidly and may be described adequately by only two- and three-body terms. Measurements of vibration-rotation tunneling (VRT) splittings for water dimers have provided data for fitting an anisotropic site potential with Woermer dispersion (ASP-W) to provide a series of highly detailed potential energy surfaces. The expressions for these surfaces include terms corresponding to electrostatic interaction, two-body exchange repulsion, two-body dispersion, and many-body induction. In this paper the authors report an investigation of the suitability of these surfaces and several others for predicting the vibrational ground-state properties of water clusters ranging from the trimer to the hexamer.

The calculations were carried out with diffusion QMC to determine cluster properties, the structures and, in particular, the vibrational average rotational constants for direct comparison with experimentally measured values. The ground-state properties were determined in runs for 1000 walkers with 15,000–20,000 time steps after equilibration. Histograms of configurations were used for calculating the internal tensors leading to the rotational constants.

Only one of the ASP-W functions and a "tuned" symmetry-adapted perturbation theory surface (SAPT5st) reproduced the rotational constants for the full range of cluster sizes up to the hexamer. These two were found to give excellent agreement with the experimental data. Investigation of the effects of adding a three-body dispersion term and consideration of many-body induction terms led to the conclusion that these small terms might be important for determining cluster properties. It was noted that the next test of the authors' VRT(ASP-W)III surface would be "to employ it in actual liquid simulations."

153

S. MORONI, N. BLINOV, & P.-N. ROY

Quantum Monte Carlo study of helium clusters doped
with nitrous oxide: Quantum solvation and rotational
dynamics

J. Chem. Phys. **121**, 3577–3581 (2004)

The infrared spectra of small molecules embedded in helium
nanodroplets display sharp rotational lines suggesting a decoupling
of the molecular rotation from the motion of the helium atoms.
The spectra observed for several different molecules have been explained in earlier reptation QMC calculations[a] giving moments of
inertia and thus effective rotational constants B as a function of
the number N of helium atoms in a cluster. This paper reports
such calculations for the nitrous oxide molecule NNO which has
been observed to have a turnaround in the evolution of B with N.
The value of B first decreases to a minimum at $N = 6$ to 7, then
increases slightly to a maximum at $N = 11$, followed by another
minimum and a slow increase.

The reptation QMC calculations provided ground-state energies, density profiles, and correlation functions for angular mo-
menta. The results gave excellent agreement with the experimentally determined values of B. The variations of B with N were reproduced quantitively, and the behavior was explained completely
with the aid of the helium density profiles showing the accumula-
tion of density at various positions surrounding the NNO molecule
as N is increased.

Additional calculations using finite-temperature path integral
Monte Carlo (PIMC) gave excellent agreement with the energetics
and structural properties given by the ground-state calculations.
The rotational constants, obtained with Boltzmann statistics in
the PIMC calculations, showed no turnaround and no oscillations,
thereby indicating that the experimentally observed behavior is the
result of exchange effects not included in the PIMC calculations.

[a]S. Moroni, A. Sarsa, S. Fantoni, K. E. Schmidt, and S. Baroni, Structure, rotational dynamics, and superfluidity of small OCS-doped He clusters,
Phys. Rev. Lett. **90**, 143401 (2003); F. Paesani, A. Viel, F. A. Gianturco,
and K. B. Whaley, Transition from molecular complex to quantum solvation
in ^4He_NOCS, *Phys. Rev. Lett.* **90**, 073401 (2003).

154

S.-I. LU

The accuracy of diffusion quantum Monte Carlo simulations in the determination of molecular equilibrium structures

J. Chem. Phys. **121**, 10365–10369 (2004)

Results of fixed-node diffusion QMC calculations for 17 small molecules made up of atoms H, C, N, O, and F were compared in this paper to experimental values for atomization energies, bond lengths, and bond angles, as well as corresponding values from coupled cluster calculations. The QMC calculations were of the Ornstein-Uhlenbeck type,[a] with trial functions for importance sampling and node locations based on linear combinations of determinants with molecular orbitals composed of floating spherical Gaussian orbitals. Initial geometries were taken from density functional calculations and optimized to find the minima given by the QMC calculations. Zero-point energies were calculated using the energy gradients given by the same calculations.

The calculated atomization energies for the 14 species for which experimental data were available were in excellent agreement with the experimental values and with values given by CCSD(T) with a cc-pVQZ basis set. The mean absolute deviation from experimental atomization energies was 0.16 kcal/mol for the QMC and 0.21 kcal/mol for the CCSD(T) calculations. Geometric parameters were in agreement with each other and with experimental values within their uncertainties. Overall, the results provide a solid confirmation of the accuracy of fixed-node diffusion QMC calculations.

[a]S.-I. Lu, A diffusion quantum Monte Carlo method based on floating spherical Gaussian orbitals and Gaussian geminals, *J. Chem. Phys.* **114**, 3898 (2001).

155

S. A. ALEXANDER & R. L. COLDWELL

A ground state potential energy surface for H_2 using Monte Carlo methods

J. Chem. Phys. **121**, 11557–11561 (2004)

Although more accurate values of ground-state electronic energies of the clamped-nucleus H_2 molecule have been obtained in analytic variational calculations with explicitly correlated wavefunctions, the Monte Carlo variational calculations reported in this paper provide a useful alternative to analytic calculations. Aside from confirming energies, they may be used in straightforward and simple independent calculations of the quantities needed for nonadiabatic diagonal corrections, relativistic corrections, and radiative corrections to the Born-Oppenheimer potential energy curve. The trial functions used were symmetric exponential functions of products of electron-electron and electron-nucleus distances with integer exponents and up to 128 adjustable parameters optimized to minimize a combination of the energy and its variance. The calculations were carried out for 24 internuclear distances in the range of 0.2 to 10.0 bohr, and they gave energies typically 1–3 microhartrees above analytic variational values with uncertainties typically less than 1 microhartree. Most interesting are the correction terms, which are not variational and were found to confirm earlier values within 1 microhartree. Binding energies calculated for several vibration-rotation levels of H_2 using the calculated potential energy curve were found in excellent agreement with those determined from spectroscopic measurements. The authors proposed that the Monte Carlo techniques used be generalized to larger systems for which analytic integrations are much more difficult.

156

J. C. GROSSMAN & L. MITÁŠ

Efficient quantum Monte Carlo energies for molecular
dynamics simulations

Phys. Rev. Lett. **94**, 056403-1/4 (2005)

This paper opened the area of molecular dynamics to "on the
fly" trajectory calculations with QMC accuracies. Using the continuous evolution of fixed-node diffusion QMC electronic structure
calculations simultaneously with the motion of nuclei the authors
were able to gain more than an order of magnitude in computational speed and make such calculations feasible for a variety of
interesting systems. "On the fly" molecular dynamics simulations
using QMC energies were thus advanced to compete with those
using density functional theory.[a]

In the simulations described the nuclear Hamiltonian and the
electronic wavefunction, represented by QMC walkers, were updated at each molecular dynamics step. This replaced discrete
QMC calculations and took advantage of correlated sampling of
electron configurations. Relaxation of the electronic wavefunction
to new nuclear positions was required, but as few as three QMC
steps per molecular dynamics step were found adequate. Calculations were carried out using nonlocal pseudopotentials with trial
wavefunctions for valence electrons consisting of Slater determinants and Jastrow functions. Variational QMC was used for optimizations.

The efficiency and accuracy of the dynamical QMC approach
was demonstrated in calculations for silicon hydride species SiH_4 to
$Si_{14}H_{20}$ at 1000 K, for the dissociation of a water molecule as H_2O
$\rightarrow$ HO + H, and for a 32-molecule liquid-water system along with
a single molecule for comparisons. The calculations were clearly
successful in producing accurate molecular dynamics simulations
at a modest cost in computational effort.

[a]R. Car and M. Parrinello, Unified approach for molecular dynamics and
density functional theory, *Phys. Rev. Lett.* **55**, 2471 (1985).

157

S.-I. LU

Theoretical study of transition state structure and reaction enthalpy of the F + H_2 → HF + H reaction by a diffusion quantum Monte Carlo approach

J. Chem. Phys. **122**, 194323-1/7 (2005)

As noted by the author the reaction F + H_2 → HF + H has played an important role in the development of theories of chemical dynamics. It has played a similarly important role in the development of accurate ab initio methods for treating the electronic structure of molecular systems, and this paper itself continues that development. After early excitements and disappointments in the 1970s and 1980s several types of quantum calculations in the 1990s were quite successful in predicting potential energy surfaces for this reaction which were entirely consistent with experimental observations. A most recent multireference coupled cluster (MRCC) study by Kállay et al.[a] provided excellent agreement. The subject paper extends fixed-node QMC calculations to provide a similarly accurate surface also in excellent agreement, solidifying mankind's understanding of this reaction.

The key to Lu's success is a multireference trial function of floating spherical Gaussian orbitals and spherical Gaussian orbitals (FSGO-SGG) suitably optimized to specify the nodes of the wavefunction. An earlier QMC calculation[b] with a single-determinant trial function produced a barrier height too high by 2 to 3 kcal/mol. The value of 1.09(16) kcal/mol obtained by Lu is in good agreement with the of 1.11 kcal/mol from the MRCC calculation. At −32.06(17) kcal/mol the QMC calculated reaction enthalpy change is in good agreement with the MRCC value (−31.94 kcal/mol) and the experimental value (−32.00 kcal/mol). The saddle point configurations (bent as suggested by the earlier QMC study) found in the MRCC and the FSGO-SGG QMC study were nearly identical. This type of quantitative agreement leads to a high degree of confidence in both methods.

[a]M. Kállay, J. Gauss, and P. G. Szalay, Analytic first derivatives for general coupled cluster and configuration models, *J. Chem. Phys.* **119**, 2991 (2003).

[b]D. R. Garmer and J. B. Anderson, Potential energies for the reaction F + H_2 → HF + H by the random walk method, *J. Chem. Phys.* **89**, 3050 (1988).

158

A. GHOSAL, C. J. UMRIGAR, H. JIANG, D. ULLMO, & H. U. BARANGER

Interaction effects in the mesoscopic regime: A quantum Monte Carlo study of irregular quantum dots

Phys. Rev. B **71**, 241306-1/4 (2005)

The paper reported the extension of earlier QMC calculations for small symmetric quantum dots to large irregular dots, i.e., from the microscopic to the mesoscopic regime. For a large number N of electrons in a sufficiently irregular confining potential the motion of the electrons may be expected to be chaotic, and mesoscopic fluctuations may result from electron interaction and interference effects. Studies using density functional theory (DFT) and the random phase approximation (RPA) have met with only limited success in treating these systems. This fixed-node diffusion QMC study was much more successful.

The systems treated were two-dimensional quantum dots containing 10–30 electrons confined by a potential energy of the type $V(x, y) = c_1 x^4 + c_2 y^4 + c_3 x^2 y^2 - c_4 (x - y) x y (x^2 + y^2)^{1/2}$. Calculations were carried out for dots formed by six different sets of constants. Each of these was investigated with each number of electrons N and the total spin S given by $S = 0$, 1, and 2 for even N and $S = 1/2$, 3/2, and 5/2 for odd N in order to determine the ground-state energies and spins. Multi-determinant trial wavefunctions with simple Jastrow functions were assembled from Kohn-Sham orbitals, and coefficients for the determinants were optimized. Both VQMC and fixed-node DQMC calculations produced energies with statistical errors small compared to single-electron energy-level spacings.

The results show a substantial probability of nontrivial spin states (about 40% with $S = 1$ for even N and about 10% with $S = 1/2$ for odd N) for the ground states, at slightly lower probabilities than given by LSDA-DFT calculations. The even-odd alternation in addition energies (similar to a second derivative) was found somewhat stronger in the DQMC results. The combination of findings was interpreted as indicating that LSDA-DFT overpredicts the effects of interactions in quantum dots.

159

A. MA, M. D. TOWLER,
N. D. DRUMMOND, & R. J. NEEDS

Scheme for adding electron-nucleus cusps to Gaussian orbitals

J. Chem. Phys. **122**, 224322-1/7 (2005)

The trial wavefunction for importance sampling in a diffusion QMC calculation for an atom or molecule has most often been assembled from a Slater determinant or a sum of determinants multiplied by a Jastrow function. The determinantal part is readily obtained from easy-to-use quantum chemistry programs which make use of Gaussian basis sets. The programs take advantage of the analytic integrals available for Gaussian exponentials, but these functions fail to give a good fit to realistic one-electron orbitals and produce large fluctuations in the local energy near the electron-nucleus cusps. One solution to the problem has been to fit the Gaussian functions to simple exponentials, adjust them to satisfy the cusp condition, and use them in the determinants.[a] In this paper a new scheme, having the advantages of an automatic procedure and avoiding some of the earlier problems is reported.

In this scheme a part of each s-type Gaussian function $\phi(r)$ is replaced by a new function $\tilde{\phi}(r)$ which depends on $\exp[\rho(r)]$ in which $\rho(r) = \alpha_0 + \alpha_1 r + \alpha_2 r^2 + \alpha_3 r^3 + \alpha_4 r^4$. The constants α_i and others are adjusted to fit first and second derivatives to $\phi(r)$, to satisfy the cusp condition, and to optimize the behavior of the local energy. This is accomplished analytically.

The scheme was investigated in detail with VQMC and fixed-node DQMC for the atom Ne and the molecule H_2. Calculations were also made for the 55 molecules of a standard test set. Improvement in the standard deviation in local energy, relative to that for no cusp correction, was dramatic. In the VQMC calculations for the test set the standard deviation in individual local energy was reduced by a factor of 2 to 5 for most molecules. This corresponds to a reduction by a factor of 4 to 25 in calculation effort for a fixed statistical error. Total energies were found to be slightly reduced.

[a] D. R. Garmer and J. B. Anderson, Quantum chemistry by random walk: Methane, *J. Chem. Phys.* **86**, 4025 (1987).

160

M. P. NIGHTINGALE & M. MOODLEY

Interdimensional degeneracies in van der Waals clusters
and quantum Monte Carlo computation of rovibrational
states

J. Chem. Phys. **123**, 014304-1/7 (2005)

In the course of an investigation of n-body correlations in trial wavefunctions for noble gas clusters the authors varied the spatial dimensionality of some of the clusters and discovered a surprising phenomenon: an interdimensional degeneracy for rotationally invariant states. For Lennard-Jones clusters of N particles in D dimensions they found identical energies for the ground states in dimensions $D = N - 1$ and $D = N + 1$. They were able to explain this behavior by considering a D-dimensional Schrödinger equation for all S-states in dimensions $D \geq N - 1$. The minimum energy was observed and proven to occur for $D = N$.

The calculations were carried out in dimensions $D = 1$ to 6 for systems corresponding to clusters of Kr, Ar, Ne, and a hypothetical $\frac{1}{2}$-Ne with half the mass of Ne. The interactions were taken as pairwise-additive L-J 6-12. Very high accuracies were obtained with the aid of trial wavefunctions which were linear combinations of basis functions with nonlinear variational parameters optimized with an iteration procedure. A typical result is that for Kr_3 in $D = 2$ and $D = 4$ producing energies of $-2.760\ 461\ 351\ 5$ and $-2.760\ 461\ 351\ 3$, respectively, in dimensionless units.

Interdimensional degeneracy was observed for hydrogen wavefunctions in the 1930s and for certain many-electron systems in the 1970s. For Lennard-Jones clusters it had not been observed previously except for the simple case of $N = 2$ in one and three dimensions. (A group-theoretical treatment for S-states had been recently reported.) As indicated by the authors a knowledge of the degeneracies provides a powerful check of the QMC calculations and estimates of accuracies. Other implications were also indicated.

161

Z. XIE, B. J. BRAAMS, & J. M. BOWMAN

Ab initio global potential-energy surface for $H_5^+ \to H_3^+ + H_2$

J. Chem. Phys. **122**, 224307-1/9 (2005)

The reactions $H_5^+ \to H_3^+ + H_2$, $H_4D^+ \to H_3^+ + HD$, $H_4D^+ \to H_2D^+ + H_2$, and $H_3^+ + HD \to H_2D^+ + H_2$ are important in the chemistry of interstellar space. Each corresponds to motion on the rather complicated nine-dimensional five-body potential energy surface of the cation H_5^+ which has permutational invariance of the H atoms. In the study reported in this paper the potential energy surface was generated not by QMC methods but by coupled cluster calculations with a large basis set, i.e., CCSD(T)/aug-cc-pVTZ. The QMC aspect of the study is in the determination of zero-point energies for the H_5^+, H_3^+, and H_2 species and their isotopic variants as well as the anharmonic dissociation energy of H_5^+.

To obtain an analytic function for the potential energy surface approximately 100,000 points on the surface were selected from classical trajectories started at the global minimum for H_5^+. An analytic function of the ten internuclear distances with full permutational symmetry was then fit to these points. The resulting surface was found in good agreement with the results of earlier high-level calculations for the ten stationary points on the surface. The zero-point energies for the several species were determined in diffusion QMC calculations for 20,000 walkers with 5000 steps for each of ten runs. For those started at the global H_5^+ minimum a stable average energy of 7210 cm^{-1} (20.6 kcal/mol) was observed in the time interval 300–7000 au. For those begun at the H_3^+ + H_2 dissociation limit an initial stable average energy of 9424 cm^{-1} (26.9 kcal/mol) was observed in the range 500–5000 au followed by descent to the 7210 cm^{-1} stable region. These rigorous zero-point energies along with a D_e value of 8.30 kcal/mol gave a dissociation energy D_o of 6.33 ± 0.03 kcal/mol, a value in good agreement with experimental values. A value for D_o obtained from harmonic estimates of the zero-point energies was in poor agreement at 5.57 kcal/mol. Further investigation suggested consideration of such zero-point energies may be necessary for classical trajectory studies.

AUTHOR INDEX

Alder, B. J.   18, 25, 30, 36, 39, 64, 102
Alexander, S. A.   74, 119, 151
An, G.   76
Anderson, J. B.   3, 10, 11, 12, 14, 15, 17, 18, 19, 24, 29, 31, 34, 39, 44, 47, 53, 55, 59, 63, 66, 71, 75, 82, 84, 89, 90, 96, 98, 116, 122, 123, 137, 153, 155
Andzelm, J.   60
Arnow, D. M.   22, 75
Aspuru-Guzik, A.   143
Astrakharchik, G. E.   136
Axilrod, B. M.   66
Aziz, R. A.   104

Baer, R.   42, 133
Bachelet, G. B.   79
Bak, K. L.   130
Bahnsen, R.   142
Balakrishnan, A.   96
Ballone, P.   85, 102
Baranger, H. U.   154
Barnett, R. N.   31, 41, 77, 80, 81, 105
Baroni, S.   117, 144, 149
Basch, H.   94, 113
Bauer, S. H.   14
Belohorec, P.   87
Berendsen, H. J. C.   124
Berne, B. J.   20, 50
Bernu, B.   93
Bettens, R. P. A.   147
Bhattacharya, A.   90
Bijl. A.   11
Bishop, R. F.   86
Blinov, N.   149
Blume, D.   136
Boghosian, B. M.   3, 71, 75, 82, 89, 90, 98
Bolton, F.   95
Bonifacic, V.   51, 111
Bowman, J. M.   120, 157
Boyd, R. J.   146
Boys, S. F.   23, 33, 37, 67
Braams, B. J.   157
Bressanini, D.   109
Broude, S.   120
Brown, D. F. R.   101
Brown, M. G.   103, 118
Brus, L. E.   63

Buch, V.   83, 118, 124
Buda, F.   145
Buendia, E.   86

Caffarel, M.   52, 60, 78, 98
Car, R.   152
Carlson, J.   58, 61, 139
Carter, C.   103
Cazzato, P.   144
Cencek, W.   24
Ceperley, D. M.   18, 25, 30, 36, 39, 48, 73, 79, 93, 102
Conroy, H.   4, 13, 74
Coulson, C. A,   146
Chebib, M.   147
Chekmarev, D.   110
Chen, B.   96, 100
Chester, G. V.   9, 27, 36, 48, 77, 117
Chiocchetti, M. G. B.   79
Chou, M. Y.   107
Christiansen, P. A.   43
Clary, D. C.   101, 103, 129
Claverie, P.   52, 60,
Clementi, E.   84
Coker, D. F.   35, 40, 45, 48, 65
Coldwell, R. L.   13, 74, 119, 122, 151
Crittenden, D. L.   147
Curtis, L. A.   134

Davidson, E. R.   146
Delaly, P.   85
DePasquale, M. F.   57, 97
Diedrich, D. L.   84, 137
Dobrosavljević, V.   62
Dolg, M.   106, 114
Doll, J. D.   20, 92, 100, 104
Drummond, N. D.   155
Dupuis, M.   33, 133

Edmiston, C.   126
El Akramine, O.   143

Fahy, S.   54, 69, 70, 125
Fantoni, S.   149
Fermi, E.   2
Feynman, R. P.   2
Filippi, C.   112, 125, 132, 145
Fink, R. F.   126
Finnila, A. B.   92
Flad, H.-J.   106, 114, 135

159

160

AUTHOR INDEX

AUTHOR INDEX

SUBJECT INDEX

activation energies 14
adiabatic diffusion Monte Carlo 115
anharmonic oscillators 60, 65, 72, 157
auxiliary field methods 42, 133
Axilrod-Teller-Muto (ATM)
    expression 48, 66

backflow 67
band gap 94
barrier heights 14, 29, 31, 53, 75, 84,
    127, 137, 153
benchmarks 21, 134, 135
biexitons 28
Bijl functions 11
Bloch equation 45
Born-Oppenheimer approximation
    13, 24, 36, 71, 82, 96, 152
Bose fluid 136

cancellation methods 15, 17, 22, 33,
    40, 65, 75, 82, 83, 84, 89, 90, 96,
    123, 130
channels 72
chromophores 145
clusters 21, 27, 35, 50, 68, 70, 83,
    85, 92, 100, 101, 103, 104, 114,
    116, 117, 118, 120, 124, 131, 132,
    144, 148, 149, 157
coalescence point 36
cohesive energy 54, 64, 79, 125
configuration interaction (CI) method
    20, 25, 37, 44, 66, 67, 87, 89, 105,
    112, 116, 126, 132
confinement 63, 136, 142
core electrons 43, 46, 56, 61, 69, 79,
    108, 111
core-valence correlation 46, 51, 108
correction methods 15, 17, 78, 123,
    151
correlated sampling 13, 38, 58, 59,
    123, 133, 152
correlation energies 24, 37, 39, 44,
    53, 55, 67, 73, 80, 85, 90, 91, 94,
    105, 107, 109, 110, 116, 122, 125,
    126, 135, 146
coupled cluster (CC) methods 86, 91,
    114, 116, 126, 130, 134, 135, 143,
    150, 153, 157
Crystal92 94

crystals 8, 28, 32, 54, 63, 64, 69, 79,
    94, 107, 125, 132, 142

damped core methods 56
density functional methods 54, 69,
    73, 79, 85, 94, 107, 127, 128, 131,
    132, 141, 142, 145, 150, 152, 154
density matrices 36, 45
derivative methods 41, 87, 122, 140
descendent weighting 9, 77, 104, 117,
    129
determinants, sparse 131, 138
difference methods 17, 30, 38, 59, 61,
    123, 133
diffusion equation 1, 2, 3, 5, 7, 18,
    39, 45, 98, 123
dipole interactions 13, 114
dipole moments 38, 81, 87, 118
Dirac equation 119
drift term 8, 17, 34, 49

effective potentials 43, 46, 51, 94, 110,
    113, 134 134, 135, 140
electron affinities 43, 46, 56, 73, 109
electron correlation 11, 23, 24, 29,
    32, 37, 44, 55, 56, 67, 80, 97, 105,
    106, 107, 112, 120, 132, 135, 138
electron gas 18, 30, 102, 141
energy derivatives 41, 87, 122, 140,
    153
Euler-Lagrange equations 125
exchange correlation 107, 141
excited states 12, 22, 35, 41, 52, 60,
    65, 68, 72, 93, 94, 114, 115, 118,
    120, 121, 128, 137, 143, 145, 146
excitons 28, 63
eximer lasers 114

fermionization 136
Feynman-Kac path integrals 52
fine structure 106
floating spherical Gaussian orbitals
    150, 153
Fokker-Planck equation 1, 2, 98

G1, G2 methods 127, 134
Gamess 94
Gaussian92 94
Gaussian functions 24, 64, 86, 131,
    138, 155

163

SUBJECT INDEX